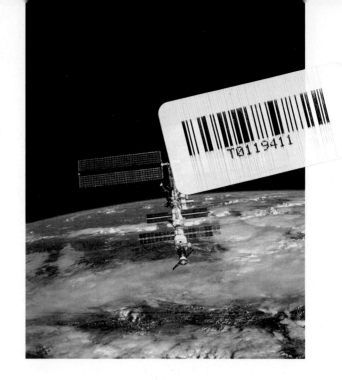

Learn about
Space and
Planets

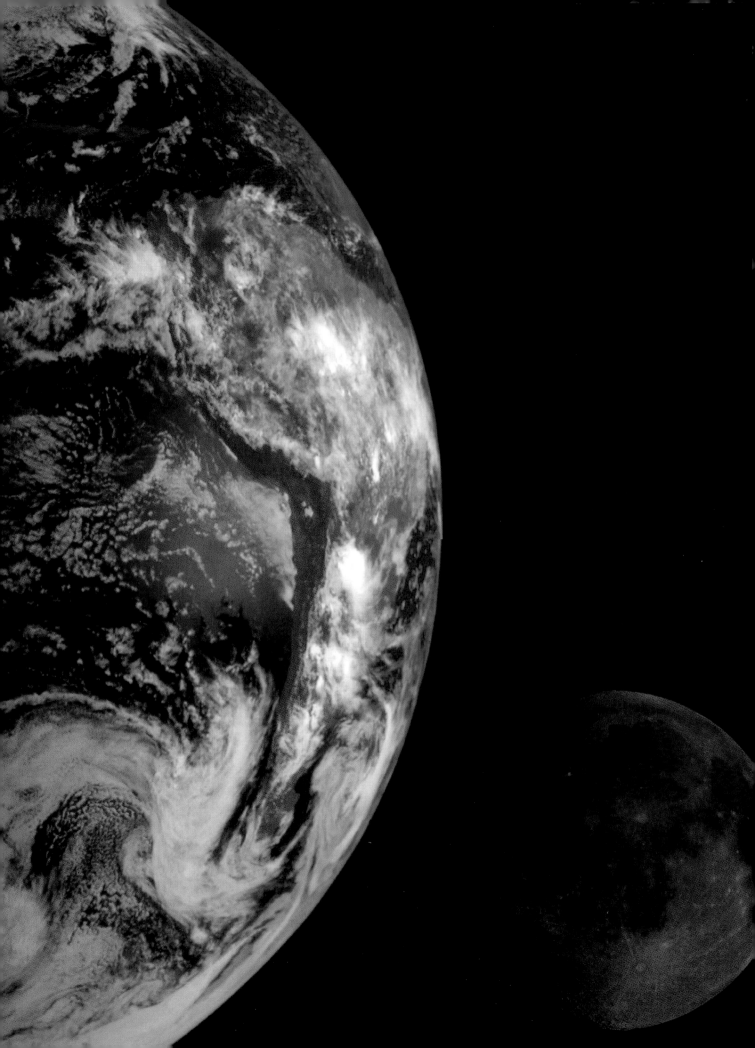

Learn about
Space and
Planets

Explore the wonders
of our universe

Susan Akass

Consultant Editor: Dr. Jacob Kegerreis (MPhys, PhD)

CICO **Kidz**

Published in 2021 by CICO Kidz
An imprint of Ryland Peters & Small Ltd
20–21 Jockey's Fields, London WC1R 4BW
341 E 116th St, New York, NY 10029

www.rylandpeters.com

10 9 8 7 6 5 4 3 2 1

Text © Susan Akass 2021
Design, illustration, and photography
© CICO Books 2021

A CIP catalog record for this book is
available from the Library of Congress
and the British Library.

ISBN: 978 1 80065 056 5

Printed in China

Editor: Caroline West
Designer: Alison Fenton
Photographer: James Gardiner
Illustrator: Stephen Dew

In-house editors: Martha Gavin
and Jenny Dye
In-house designer: Eliana Holder
Art director: Sally Powell
Head of production: Patricia Harrington
Publishing manager: Penny Craig
Publisher: Cindy Richards

Contents

Introduction

The mystery of space

Have you ever been outside on a cloudless night, somewhere with no electric lights and where the only light is from the Moon and the stars? When you look at the sky in a place like that, it is hard to believe how many stars you are able to see. You can see even more if there is no Moon shining. That's because most stars are so faint you can only see them in total darkness.

In the past, when there were no electric lights, people saw the stars every cloudless night. People wondered what stars were and why the Moon changed shape and position, and when the next full moon would be. They also wondered why the Sun traveled across the sky during the day, why days were sometimes short and sometimes long, and why there were seasons. They wanted to know when to plant their crops or when the next rains would come.

At first, people answered these questions by telling stories. The ancient Greeks thought that the Sun was the god Apollo. Every day he rode a chariot pulled by fiery horses across the sky to bring light to the world. For the Egyptians, it was the Sun god, Ra, who sailed a ship across the sky. Each night he sailed it through the underworld, where the dead go, and then he was reborn each morning. There were stories to explain everything, from seasons, to eclipses, to how the world was created.

Star watchers

Then people began to study the stars. They watched and recorded how the stars and the Moon changed over time. They realized that there were patterns: that there were about 29 days from one full moon to the next (that became the length of our month) and that the patterns of the stars and the position of the Sun took 365 days to come back to where they started (that became our year). About 12 months fitted into each year but not exactly, so our calendar has been tweaked through history with different numbers of days in a month and leap years to make it all add up.

These star watchers, some living thousands of years ago, were incredibly clever and were even able to work out when solar eclipses would happen and the world would go dark (which was terrifying for them). But they believed that Earth was the center of the universe and that the Sun and the planets revolved around it. They also believed that the stars and planets had magic powers to control our lives.

Some of these star watchers were much more scientific and didn't believe in this magic power. They were known as astronomers. They wanted to know how the universe worked. Copernicus was the astronomer who eventually pointed out that the idea of everything revolving around Earth just didn't fit with what he could see happening in space. It was the Sun that was at the center of our Solar System.

Then an Italian scientist called Galileo Galilei built a telescope to study the stars. For the first time Jupiter's moons were visible from Earth. He used his observations to show that Copernicus was right and the planets did revolve around the Sun. Later, an English scientist, Isaac Newton, came up with important ideas about forces and how things move, including how gravity keeps planets in orbit around the Sun. Modern astronomy began

to develop and, as the questions kept getting harder to answer, so telescopes got bigger and more complicated.

People became more and more interested in the idea of space travel and in 1957 the first rocket was launched to send a satellite into space. Shortly afterward, a Russian dog, named Laika, took off in a rocket and orbited Earth. Then, in 1969, two men walked on the Moon. Unmanned space travel has developed hugely since then and spacecraft have been sent way out to the edges of our Solar System to send back information. Also, for the past 20 years the International Space Station has been orbiting Earth with scientists onboard learning about space and what lies beyond our planet. Computer technology has also helped to make sense of all the information that is beaming back to Earth.

Some discoveries and ideas about space are very difficult to understand, but the best place to begin is at the beginning and to think about all the questions people first asked about our planet and the stars beyond. This book will try to answer these questions and uses fun, practical activities to help you understand some of the science behind the answers.

Become an astronomer

Step 1
Go outside at various times on a sunny day and think about these questions. Never look directly at the Sun—you will damage your eyes.

* **What is the Sun?**
* **Why does the Sun change position in the sky?**
* **What time does the Sun rise/set? Why does this change through the year?**
* **At sunset, how long does it take for the Sun to disappear?**
* **In what compass direction does the Sun rise/set?**
* **How does a compass work?**
* **How hot is it? Why is the Sun hot?**
* **What season is it? Why do seasons change?**
* **Why do we get wind and clouds?**
* **Why is the sky blue?**
* **What is above your head if you keep going up?**
* **What is beneath your feet if you keep going down?**
* **Why do things fall when you drop them?**
* **Why is there life on planet Earth?**
* **Do aliens exist?**

Step 2
On a cloudless night, ask an adult if you can go outside and look at the sky. This is easier in the winter when it gets dark earlier. Take a flashlight (torch), so you don't trip. If you have to go beyond your backyard, you will need an adult to go with you. Get as far away from streetlights or other lights as you can. Stay outside for at least half an hour to let your eyes get used to the dark and keep watching the sky. Use binoculars or a telescope if you have them. Think about the following questions:

* **Why do we get day and night?**
* **Can you see the Moon? Why does the Moon change shape?**
* **What are the shadows on the Moon?**
* **Why can't we always see the Moon?**
* **Why doesn't the Moon fall down to Earth?**
* **How many stars can you see?**
* **How can you tell if a star is really a planet?**
* **Is it possible to spot all the planets?**
* **How far away are stars?**
* **Can you see patterns of stars in the sky?**
* **Why are some stars brighter and some different colors?**
* **Why is there a long patch of mistiness across the sky?**
* **Did you see a shooting star? What is a shooting star?**
* **Is it possible to see a black hole?**

Now use the book and activities to find the answers to these questions!

Chapter 1
The Sun and Earth

THE QUESTIONS THIS CHAPTER ANSWERS:
What is the Sun?
Why do we get day and night?
Why does the Sun change position in the sky?
At sunset, how long does it take for the Sun to disappear?
In which compass direction does the Sun rise and set?

What is the Sun?

The Sun is our own personal star. It is like billions of other stars in the universe, but it is the only one close enough for us to feel its heat. All the other stars are much too far away for us to see as anything more than pin pricks of light in the sky. But what is a star? It is a massive ball of hydrogen gas which gets squeezed together at its center to make a new gas called helium (the gas you get in birthday balloons). This process is called nuclear fusion and it releases huge amounts of energy into space. Some of the energy we see as light, some we feel as heat, and some is other forms of radiation that scientists have found out about (see page 92).

We can start to use words like huge, massive, or vast for the size of the Sun and the amounts of energy it throws out into space, but the numbers are so big that the words don't help us imagine it. A million Earths could fit into the Sun. The temperature at its center is 27 million °F (15 million °C). The Sun is so distant (150 million kilometers/93 million miles away) that it takes light eight minutes to travel from the Sun to Earth.

Don't worry too much about the numbers. It's more important to understand how the Sun affects our Solar System. Earth is one of eight roughly spherical (ball-shaped) planets which, together with countless other small objects, circle the Sun. They are held in these paths, called orbits, by the Sun's gravity (the force that pulls objects toward each other). Because it is so massive, the Sun has an enormous power to hold everything in orbit around it (see page 80).

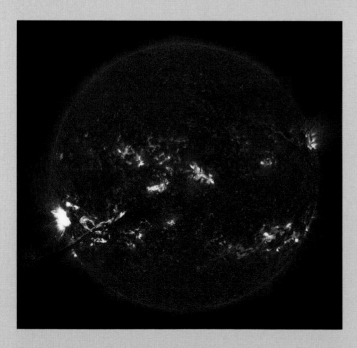

The Sun is our own personal star—a massive ball of exploding gases whose temperature reaches 27 million °F (15 million °C) at its center.

Why do we get day and night?

Everything in the universe is turning. Planets move around the Sun, but each of the planets and the Sun itself also spin around an axis. An axis is an imaginary stick that goes through the middle, like a wire through a round bead. The ends of the stick are called the poles. On Earth these are called the North Pole and the South Pole. The widest part in the middle of the ball is called the Equator. The Sun spins around its axis about once every 26 days but as we can't look at it without special equipment, we can't easily see it turning!

More importantly for us, Earth spins quite fast on its axis. When your part of Earth turns to face the Sun, that is your daytime. It is light and warm and plants can grow. When it turns away, darkness falls and you have night. Without light and warmth most plants can't grow and most animals can't live. For life, the fast turn of Earth is important.

A long time ago, astronomers decided to divide the time it takes for Earth to turn once on its axis into 24 hours—they could easily have decided on another number. In the past, it didn't matter much how many hours there were in a day because people didn't have clocks. They knew it was the middle of the day when the Sun was at its highest point in the sky. Bedtime was when it got dark because most people couldn't afford candles.

If you keep checking the position of the Sun in the sky, it may seem as if the Sun is moving, but it's not. It is you that is moving because you are standing on the spinning Earth. It's like looking out of a car window. Everything is flashing past. It seems to be moving, but it isn't.

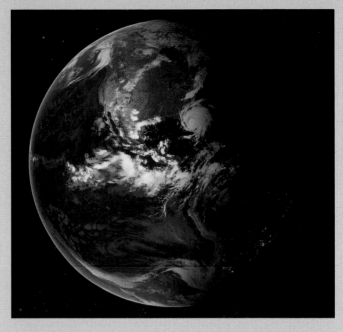

Looking at Earth from space, you can see the clear divide between light and darkness, day and night.

Everything in the car seems to be still because it's all moving at the same speed as you are. Anyone standing on Earth's equator is actually moving at 1,600 kilometers per hour (1,000 miles per hour). You can't tell you're moving because Earth is so enormous and everything around you is moving at the same speed as you. If you watch a sunset, the Sun seems to disappear from view quite quickly. That gives you some idea of how fast Earth is turning.

If you were to visit other planets, you would still have day and night, but they could be much longer or shorter than our days on Earth. It all depends on how quickly the planet turns on its axis. Mercury takes 1,408 hours to turn on its axis (nearly 59 earth days) whereas Jupiter zooms round in only ten hours!

What's a Million?

You will need

30 sheets of letter size (A4) graph paper marked with 1mm squares

Pencil

Ruler

Sticky tape

Scissors

When we talk about space we toss around words like millions and billions without really understanding what they mean. This activity will help you understand just how big a million is. A team of people cutting out makes this activity quicker. Get some friends or your brothers and sisters to help!

1 Look carefully at the graph paper and you will see that it is marked with tiny squares, but also bigger squares outlined in bold. Each side of the bigger squares is ten tiny squares long. Ten rows of 10 squares makes a hundred, so there are a hundred tiny squares in each bold square.
10 x 10 = 100

2 Now count along a row of ten of the bold squares and draw a line at the end. In this row you have ten lots of a hundred tiny squares, which is a thousand tiny squares.
10 x 100 = 1,000

3 Count down ten rows and draw a line under the tenth row. You now have a bigger square. Ten rows of a thousand tiny squares is ten thousand tiny squares. Cut out the bigger square.
10 x 1,000 = 10,000

4 Now cut out ten of these big squares and stick them together with sticky tape to make a long row. Ten times ten thousand is a hundred thousand tiny squares.
10 x 10,000 = 100,000

5 Now if you are patient and determined enough, make nine more rows like this. Stick them all together to make one huge square. Ten times one hundred thousand is one million!

10 x 100,000 = 1,000,000

You have put together a million tiny squares!

Now imagine each of those tiny squares is as big as Earth but all squished together into a ball. That's how big the Sun is.

You would need a thousand of these million-square sheets to make a billion. Line them up and the line would be one kilometer long (about two-thirds of a mile)!

Make a Sundial

Sundials were the first clocks. On a sundial, a stick called a gnomon blocks the rays of the Sun (which have traveled 150 million kilometers/93 million miles to Earth) and makes a shadow line. As Earth rotates (and the Sun appears to move across the sky), light hits the gnomon from different angles and the shadow changes direction and length. Mark the shadows and you have a clock!

1 Carefully make a small hole through the center of the plate with the point of your scissors.

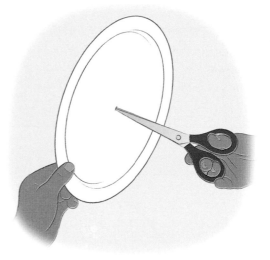

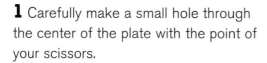

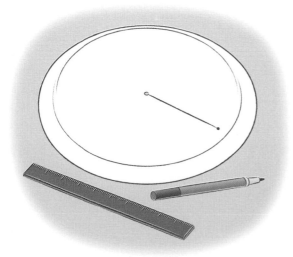

2 Turn the plate upside down. Your sundial will be on the back of the plate. Make a dot on the outside edge with the marker pen. Use the ruler to join the dot to the center of the plate and draw a line. Write 12 noon on this line.

3 Put a small amount of modeling clay under the center of the plate and push the straw through the center hole into the clay. Then mold more clay around the straw on the other side to hold it firmly upright. You don't want it to move.

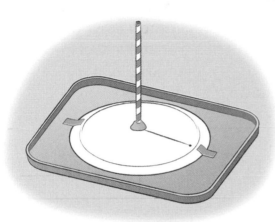

4 Use sticky tack to stick the plate firmly to the board/tray and also some sticky tape at the end of the 12-noon line and on the opposite side. This will help to keep the plate from moving or blowing away. Don't put sticky tape around the plate or you won't be able to mark it.

5 Just before 12 noon take the board into your backyard. Ask an adult where it will stay sunny all day and not get shadowed by houses, trees, or fences. At exactly 12 noon, put the board on the ground. The straw will make a shadow on the plate. Turn the board until the shadow lines up exactly with the line on the plate. Now set an alarm for 12.55 to be ready for 1pm. You can go off and play for an hour!

6 When the alarm goes off, run outside again and, at exactly 1pm, check the shadow. It will have moved and will be a little longer. Mark the end of the shadow with the pen. Mark another point on the shadow, closer to the center. Use your ruler to join these points with a line along the shadow toward the center but be careful not to knock the straw or move the plate or board. Label this line 1pm. Set your alarm for 1.55 and repeat. Keep going, marking the shadow every hour until it gets dark. Leave it overnight.

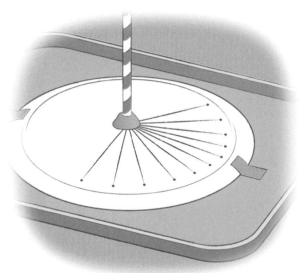

7 In the morning, when it is light, mark some more hours. You will notice that the shadows are on the other side of the 12-noon line now.

8 As long as you keep the 12-noon shadow in line with the 12-noon line on your sundial, your sundial will be accurate every day—but it won't tell you what time it is at night!

Day and Night

It's easy to talk about Earth turning on its axis, but difficult to imagine. What you need is a model, so you can see how it works. If you have a globe, this is a good opportunity to use it.

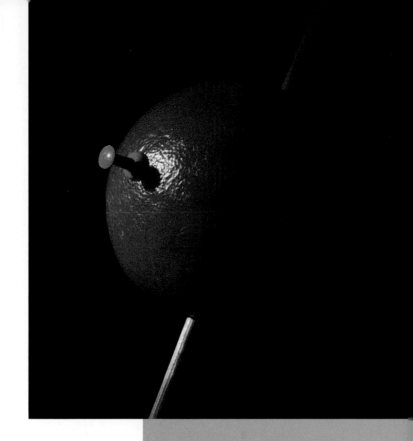

1 Either do this activity after dark or find a dark place. A bathroom with no windows works well or you could make a den with rugs under a table!

2 If you are using the modeling clay or an orange, very carefully push the bamboo skewer right through the middle to make the axis which Earth spins around.

3 Push the thumbtack into the Equator of your model Earth. This is like a giant standing on the surface!

You will need

Ball of modeling clay or an orange, as a model Earth (or a globe if you have one)

Bamboo skewer

Thumbtack or push pin with long head

Camping lantern or flashlight (torch), as a model Sun

Old cardboard box or pile of books

Compass or smart phone (optional)

Sticky tack (if you are using a globe)

4 Set up the flashlight or lantern in your dark place. Put the flashlight on the cardboard box or a pile of books.

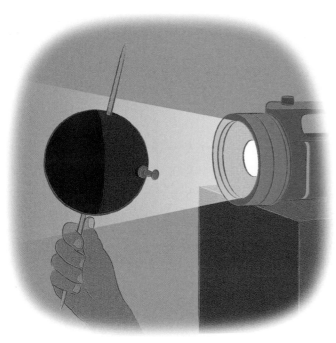

5 Use the skewer to hold your Earth in the light from the flashlight. Can you see that the side nearest the Sun is lit up and the side away from it is in shadow? There is a fairly clear line between the light and shadow. Earth's axis is slightly tipped over, so angle yours a bit to make it more realistic (tip it so the skewer points to about the number one on a clock face)—the globe will already be tipped.

6 Slowly twist the skewer (Earth's axis) so Earth turns. Watch the thumbtack move from light to darkness and back into the light.

7 Which way did you turn your Earth? If you used a compass to find out which direction the Sun is in when it rises and sets (see page 8), you will have found out that the Sun rises in the east (or on the right if you are facing north) and sets in the west. You will have to turn your Earth clockwise (from left to right) for your giant thumbtack to see the Sun rise in the east.

The planet Venus turns in the other direction, so the Sun rises in the west and sets in the east on Venus.

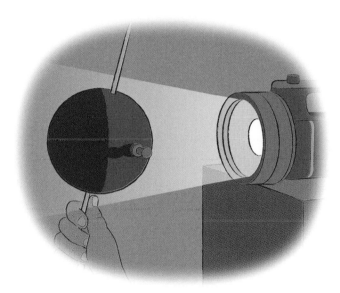

8 Now look very carefully at the shadow made by the thumbtack. Can you see how it changes direction as Earth turns. When the Sun is right overhead the shadow is tiny, but it is very long as the Sun rises or sets, just like the shadow you made on your sundial (see page 16).

9 If you are using a proper globe, stick a ball of sticky tack on your home country. When your home country is in daylight, what is happening to countries on the opposite side of the globe? They are in darkness. For instance, when it is 12 noon in London, UK, it is about 10pm in Sydney, Australia. When they are in daylight, you are in darkness. If you want to phone a friend on the other side of the world, you must be careful not to phone in the middle of the night! Check out online what time it is in different cities in the world compared to your time. This is a good website to use: https://24timezones.com/#/map.

Chapter 2
The Solar System

THE QUESTIONS THIS CHAPTER ANSWERS:
How can you tell if a star is really a planet?
Is it possible to spot all the planets?

– – – – – – – – – – – – – – – – – – – –

Stars and planets

Our Sun and everything that spins around it is called the Solar System. The biggest things that are spinning (orbiting) around the Sun are the planets.

There are eight planets, which are all very different. The four nearest the Sun are Mercury, Venus, Earth, and Mars and these are called the rocky planets because they are made mostly of rock. Mercury is the smallest planet; Venus, which is covered with a thick layer of cloud, is the hottest. Mars is a desert planet and the one we know most about (other than Earth) because several unmanned spacecraft have landed there and sent back a lot of information and images. We know that there used to be liquid water on the surface of Mars and there might have been life there once (or maybe it still exists below the surface—see page 114!)

The next two planets are the gas giants, Jupiter and Saturn. No spacecraft will ever land on them because they are mostly made of liquid and gas with no surface to land on. But you could walk on their many moons, which are rocky like the inner planets. Jupiter is by far the biggest planet. If you pulled the stuff that all the other planets are made of into a ball, it would still be less than half of what makes up Jupiter. Saturn is spectacular because of the beautiful icy rings around its equator.

Beyond the gas giants are the ice giants, Uranus and Neptune. These are mostly made of ice and rock and, like the gas giants, do not have a surface you could land on. The ice is made from frozen water and two frozen gases, ammonia and methane. It is the methane (see page 78) that makes these planets look blue. The most interesting thing about Uranus is that

it rotates on its side and looks as if it is rolling round its orbit. Neptune is a cold, stormy world, 30 times farther from the Sun than Earth.

Spacecraft have visited or flown past all the planets and sent back wonderful photos and lots of information about them. You can find out some of the details in the reference pages at the back of the book.

Between Mars and Jupiter is the asteroid belt where millions of pieces of rock, some huge, some tiny, orbit the Sun. It sounds as if it should be crowded but space is so big that there is about a million miles between each of them! One of these asteroids, called Ceres, is big enough to be called a dwarf planet.

Beyond Neptune is the Kuiper Belt where there are hundreds of dwarf planets—planets smaller than Mercury. One of these, Pluto, used to be called the ninth planet but then scientists changed their minds about what makes a planet. Astronomers are discovering more and more dwarf planets in this belt.

The order of the planets from the Sun is Mercury Venus Earth Mars Jupiter Saturn

You cannot put a scale diagram (see page 24) of the Solar System in a book. If the Sun was small enough to go in the picture, the planets would be too small to see, and the distances between them are so huge that they would be off the page! This diagram shows the order of the planets and roughly how big they are compared to each other (but not compared to the Sun).

Uranus Neptune. To remember the order, learn one of these helpful mnemonics (where the first letter of each word is the first letter of the planet):

**My Very Easy Method
Just Speeds Up Names**

or if you prefer

**My Very Excellent Mother
Just Serves Up Noodles!**

Spot Planets

If you are really interested in astronomy, you will want to spot the planets for yourself. Mercury, Venus, Mars, Jupiter, and Saturn are bright enough for most people to be able to see without a telescope. Mercury is hard to spot because it is nearest to the Sun. You have to look for it close to the rising or setting Sun, but the glare from the Sun makes it hard to see.

You will need

Use of the Internet to find the best time for spotting planets

A cloudless sky and a new moon (see page 50)

An adult to take you somewhere dark away from streetlights

Suitable clothes (it can get cold after dark)

Compass

Flashlight (torch) to help you move around in the dark

Binoculars or a telescope (optional)

1 To help you know when and where to look, use a website (such as https://www.timeanddate.com/astronomy/night/). You will need to enter the date and where you live and then the website will tell you which planets are visible, at what time, and in which compass direction.

You can often spot planets before it gets light, just before dawn, or after it gets dark, just after sunset—you don't need to be out in the middle of the night! This is easier in the winter when it gets light later and dark earlier.

2 Stars twinkle, but planets don't, so look for big, bright stars that don't twinkle. Stars are suns, so they make their own light, but they are so far away that we can only see a tiny point of light. That light is bounced around by Earth's atmosphere and that makes them twinkle.

Planets don't make their own light. You can only see them because they reflect sunlight, like the Moon. Their light also gets bounced around by Earth's atmosphere, but because they are bigger you can't see it happening as much. They look bigger because they are much closer to Earth than stars. Use binoculars or a telescope and they will help you to see the planets more clearly, but you don't need these to identify them.

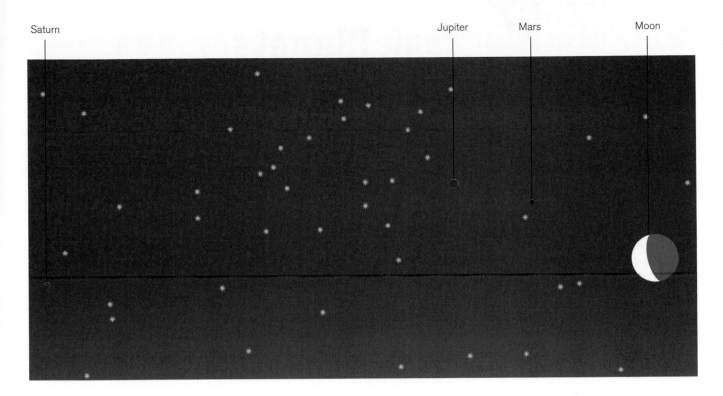

Saturn   Jupiter   Mars   Moon

3 Color helps you identify which planet is which. Mercury is whiteish, Venus is bright white, Mars is a rusty orange colour, Jupiter is a light tan. If you have good binoculars, you might see four bright points of light near Jupiter which are four of its moons. You need a telescope to see Saturn's rings, but you can see its golden color through binoculars.

4 When you are searching for planets, you may be lucky enough to see the International Space Station tracking across the sky, reflecting the Sun just like a planet. Because it moves fast (it will only be visible for between one to five minutes), you may think it is a plane, but it moves steadily without any flashing lights. It is the biggest man-made object in space—the size of a soccer (football) pitch. You can work out when it is visible by using a website (try https://www.accuweather.com/en/space-news/how-to-see-the-international-space-station-from-your-backyard/348602). The Hubble Space Telescope is also easy to see and if you are in a really dark place, you may spot other satellites, too.

A Scale Model of the Solar System

Making models of the Solar System is fun and will help you remember the order of the planets and what they are like. It will also help you understand just how huge the Solar System is. But it's not easy to make a scale model. A scale model means that even though it is much smaller, everything still looks right. If you made a model of yourself, you wouldn't make your body a quarter the proper size and your head half the proper size because your head would look much too big. Everything in a scale model must be made smaller by the same amount.

This model uses things from around the house. The measurements are in millimeters as these are the easiest to use for very small measurements. This model uses the diameter of the planets—that is the distance through the middle from one side to the other.

Sun

Mercury   Venus   Earth   Mars   Jupiter   Saturn   Uranus   Neptune

Ruler marked with centimeters and millimeters

Permission to use a few things from the fridge, store cupboard, and backyard!

1 Find a soccer ball (football) or basketball. This ball will be the Sun in your model Solar System. It has a diameter of about 230mm.

2 Using the same scale, Mercury is smaller than 1mm. A pinhead is about the right size to be Mercury, but that has a pin attached! So, use a mustard seed if you can find one or maybe a grain of sand.

3 Venus and Earth are about the same size as each other, at around 2mm. Use two round peppercorns for these.

4 Mars is a bit bigger than Mercury. Just a bit more than a millimeter. Use another mustard seed or grain of sand.

5 Now things are getting bigger: Jupiter is nearly 24mm. A largish cherry tomato is about right! Put it on a ruler to check the size.

6 Saturn is a bit smaller at 19mm. A fairly big grape or large marble would do.

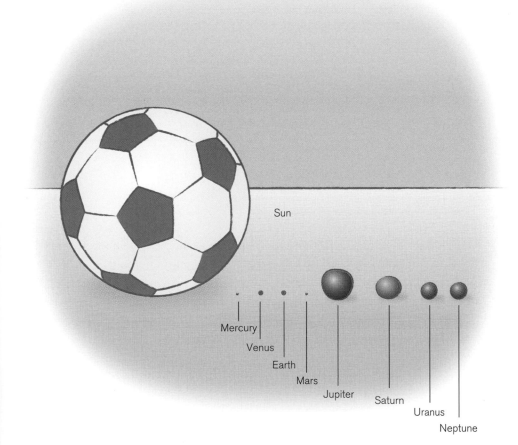

Sun

Mercury

Venus

Earth

Mars

Jupiter

Saturn

Uranus

Neptune

7 Uranus and Neptune are about the same size as each other, at nearly 8mm. An ordinary marble is about right. Two chocolate Maltesers are a bit big but would do!

8 If you wanted to include Pluto, a dwarf planet, it is smaller than half the size of Mercury. Use a grain of fine salt.

Now you know how small planets are compared to the Sun!

Painting Planets

You can begin to understand more about planets by painting them, but to do this we have to forget about using an accurate scale because the Sun and Jupiter are too big to paint and Mercury too small. Use the pictures on pages 30 and 31 to help as a guide when you paint each planet.

1 Mercury is the smallest of the rocky planets. Its gray rocky surface is pitted with craters a bit like our Moon.

For Mercury, use a pencil to draw around a beaker on a small piece of paper to make a small circle. Cut out the circle and place it on the newspaper to paint. Put a small squirt of white paint and a small squirt of black paint on your mixing plate. Swirl them around a bit but leave the paint streaky (not mixed). Paint Mercury thickly, making sure that there are streaks of darker and lighter gray. Use a sponge to make a blobby, crater-covered surface.

2 Next, use a saucer for Venus, so that it is bigger than Mercury. Venus is completely covered with clouds so you can't see its rocky surface.

Mix some yellow, black, and white paint to make a pale grayish tan color and paint Venus. Then dip a big dry paintbrush in some dark brown paint and dab on some darker patches. Dip another dry brush in white and drag it across the surface in a curve to make drifts of cloud.

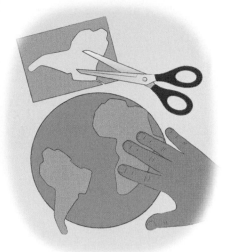

3 For Earth, draw another circle the same size as Venus. Earth stands out in the Solar System because so much of it is covered with water.

Paint Earth bright blue for the oceans and let it dry. Cut out continents from green paper and stick them on—they don't need to be accurate! Swirl on clouds using a thick dry brush and white paint.

4 Make Mars a little bigger than Mercury—use a bigger beaker. It is a dry, rocky planet with only a very thin atmosphere.

Paint Mars all over in an orangey brown and then use a sponge dipped in dark brown to make shadowy patches.

5 Jupiter is the largest planet—a gas giant with no solid surface. Jupiter looks stripy because the gases move in different directions. Light bands are where gas is warmer and is rising. Dark stripes are where the gas is colder and falling (see page 56). Where the gases moving in different directions bump into each other, you get swirls of gas and huge storms.

Find your biggest sheet of paper and make the biggest circle you can—use a mixing bowl or bucket! Paint Jupiter with curved stripes in yellows, white, cream, and pale browns using thick paint. While the paint is still wet, use a dry brush to swirl and blend the stripes together. In one of the lower stripes make a spiral of darker reddish orange—that's Jupiter's great storm (see page 57).

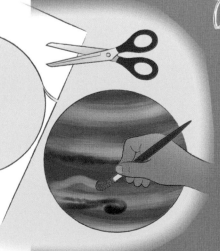

You will need

3 large sheets of white paper (ledger size/A3 or bigger) or the side of a big cardboard box

Letter size (A4) white paper

Pencil

Beakers, saucers, plates, and mixing bowls to draw around (ask permission first)

Scissors

Newspaper to protect the countertop

Acrylic paint set

Mixing plate

Big paintbrushes

Small sponges

Green paper

Silver marker pen

Sticky tack to display your planets

6 Saturn is big but not as big as Jupiter and stands out because of its beautiful rings. These are made of millions of pieces of ice, some huge, some tiny, which are all orbiting the planet in a layer that is only a few meters thick.

Draw around a big dinner plate for Saturn, but don't cut it out because you need to draw the rings on after it is painted. Saturn is stripy but not as stripy as Jupiter. Use wider stripes of pale grayish yellow and grayish blue blended in the same way as you did for Jupiter. Let the paint dry before you draw on the rings. Practice drawing the rings on some scrap paper first to feel how the curve works before you draw them on your planet. Use lines of silver pen to color them. If your wall is white you could leave the white paper in between the rings and the planet, or if you want to, use pointy scissors to cut these inside pieces out.

Sun

Venus

Mercury

Earth

Mars

7 Uranus and Neptune, the ice giants, are similar in size, but smaller than Saturn and much bigger than Earth. They are both blue because of methane gas but they are different shades of blue.

Find a plate or bowl to draw around and draw two circles. Paint Uranus smoothly in a very pale aqua. Paint Neptune in a brighter blue with stripes of darker blue, using the same dry brush method you used for Jupiter.

8 Finally, make the Sun. You won't be able to make a Sun which is nearly big enough, so just make a thin section of the Sun to put at the edge of your display. Draw a slightly curved line at the edge of the biggest piece of paper/card you have. Start the curve at the top-left corner and end it at the bottom-left corner. Paint it in thick swirls of red, orange, and yellow paint.

9 Fix your Sun with its flat edge in a corner of a room and then display your planets in the correct order (see page 21) on the wall next to it.

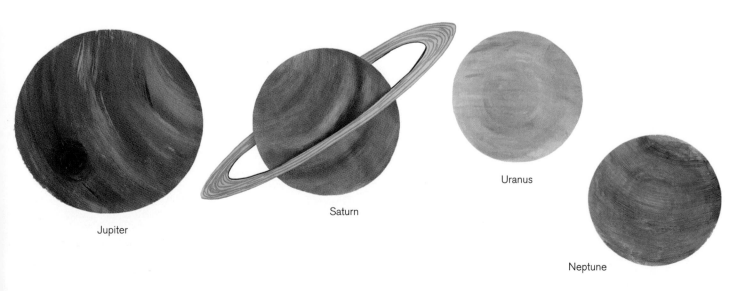

Jupiter

Saturn

Uranus

Neptune

Chapter 3
Orbits Around the Sun

THE QUESTIONS THIS CHAPTER ANSWERS:
Why doesn't the Sun rise and set at the same time all through the year?
Why do the seasons change?

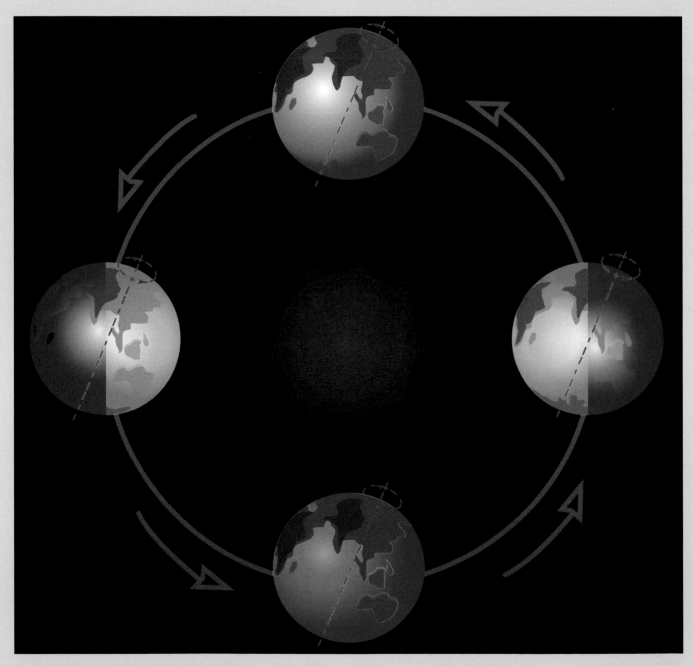

Elliptical orbits

Everything in the Solar System orbits around the Sun in what seems like a fixed pattern, although it has changed sometimes billions of years in the past. If the orbits were circles, then everything would always stay the same distance from the Sun—like the valve on your bike tire is always the same distance from the hub at the center—but they're not. The orbits are elliptical, that is an oval shape. The planets' orbits are not far-off circular, but Pluto's orbit is much more elliptical and comets have really stretched-out orbits! This means that sometimes they are much closer to the Sun than at other times. The planets travel at different speeds in their orbits, so, seen from Earth, they seem to catch each other up and sometimes they are closer to Earth than at other times (and will be brighter in the sky).

It is hard to understand just how big distances in space really are. When scientists talk about the distance of a planet from the Sun, they measure it in AU, or astronomical units. One AU is the distance from Earth to the Sun, which is about 150 million kilometers (93 million miles). Using AU makes the numbers smaller and easier to work with. Neptune, the farthest planet from the Sun, is 4,490,000,000 kilometers (2,790,000,000 miles) from the Sun, which is 30 AU. Mercury, the closest planet to the Sun, is less than half an AU from the Sun.

Earth years and seasons

From way back in history people have divided time into years as well as days, and that is because of how long it takes the Earth to orbit around the Sun. Astronomers in the past worked out that a year lasts 365¼ days. We

LEFT: The seasons on Earth are caused by the tilt of its axis. When your part of the world is tilted away from the Sun, it is your winter; when it is tilted toward it, it is your summer.

Uranus's orbit lasts 84 Earth years. Because Uranus is tilted almost onto its side, a very wide area of each hemisphere has 21 Earth years of darkness in its winter and 21 Earth years of daylight in its summer.

have a leap year of 366 days every four years because of the extra ¼ day in each year.

Other planets have very different lengths for their years. The closer they are to the Sun, the shorter their years. This is because their orbits are shorter—like being on the inside lane of a running track. Being close to the Sun also makes these planets travel faster because the Sun's gravity is stronger (see page 80). Mercury's year is only 88 earth days long, whereas Neptune's year is 60,190 earth days. That's a long time to wait for a birthday!

If you think about a year, you probably think about the changing seasons. The reason we have changing seasons is because our Earth is tilted on its axis and that means sometimes parts get more sun than at other times of the year. Planets have different amounts of tilt. Mercury, Venus, and Jupiter are hardly tilted at all. Uranus is right over on its side and seems to roll around its orbit. The others are tilted in a similar way to Earth.

Uranus

Saturn

Jupiter

Mars

Earth

Venus

Mercury

Book

Sun

How Far Away Are the Planets?

Remember the basketball and peppercorn models of the planets we used in the scale model activity on page 24? Let's keep to the same scale and try to place the planets in a row at the correct distance from the basketball-sized Sun.

Put your Sun down in your backyard and take ten big, one-yard/meter paces, then put down the mustard seed that is Mercury. Take another nine paces and put down the peppercorn Venus and then another seven for the peppercorn Earth. You will need another 14 paces before you can put down the mustard seed Mars. You will have lost sight of the other three planets by now! You have to take another 95 paces before you can put down Jupiter and will probably be well outside your backyard. If you carried on, Neptune, the planet farthest from the Sun, would be 642 paces beyond Jupiter. Space is big!

Let's change this model a bit and use the unit AU to show the distance of each planet from the Sun. This makes the spacing easier to understand.

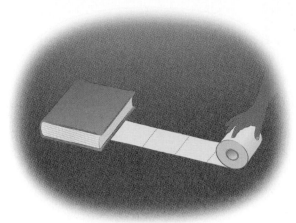

Heavy book to use as
a weight

Roll of toilet paper

Colored modeling clay

1 At the end of a corridor or a big room, or in your backyard, put a book on the loose end of the toilet roll, covering the whole of the first sheet, to hold it down. The book is the Sun.

2 Let's say one sheet of toilet paper stands for 1AU. Earth is 1AU away from the Sun, so put a small ball of modeling clay on the next join between the sheets. Mercury and Venus come between Earth and the Sun. Put a tiny lump of clay close to the Sun for Mercury and a lump about the same size as Earth, between Mercury and Earth, for Venus.

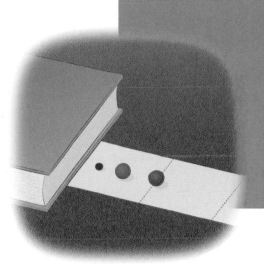

3 Use another tiny ball for Mars, half a sheet on from your Earth.

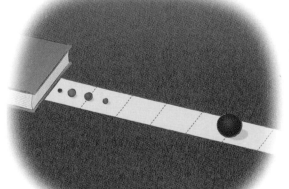

4 Make a really big ball for Jupiter. Unroll the toilet roll a little and count five sheets from the Sun before placing it down.

5 A big ball for Saturn goes nine-and-a-half sheets from the Sun.

6 A smaller ball for Uranus is 19 sheets from the Sun.

7 Neptune is about the same size as Uranus and is 30 sheets from the Sun!

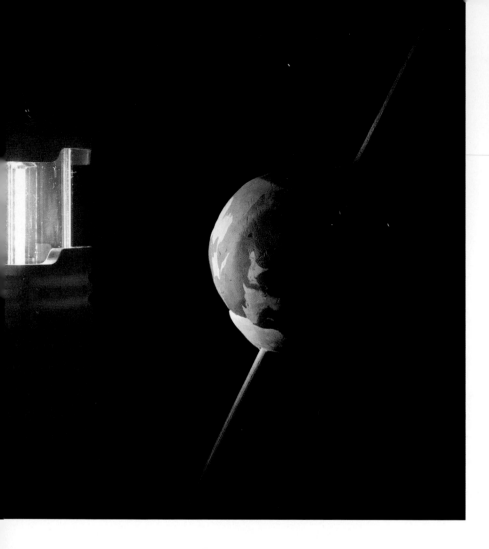

A Planetary Year

Why seasons happen is easier to understand if you use a model. This activity is a great place to use a proper globe if you have one. Otherwise, a large ball of modeling clay will work fine.

1 If you are using modeling clay, make it into a ball for Earth. Push the bamboo skewer through the middle of the ball like an axis, so it comes out at the North and South Poles. (Remember there is no real pole there!) Make two small balls of white clay. Thread them onto the skewer at the North and South Poles, then flatten them to make the ice caps.

You will need

Globe or large ball of colored modeling clay

White modeling clay

Bamboo skewer (if using modeling clay)

Dark room

Camping lantern or an electric lamp with no shade

Thumbtack or push pin (if using clay) or some sticky tack (if using a globe)

2 Now you need to find a dark place to set up your lantern Sun—the darker, the better. The only light should come from the lantern. If you are using an electric lamp, ask an adult to help you set it up. It is important that the light comes from the lantern or lamp in all directions, so a flashlight (torch) won't work. You also need space for your Earth model to orbit around the Sun.

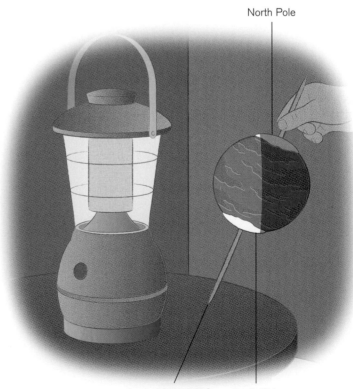

North Pole

3 Switch off all other lights and switch on the Sun. Hold the model Earth by the skewer so it is tilted as if it were pointing to the number one on a clock face. (The globe will be tilted already.) Point the North Pole so that it is tilted away from the Sun and pointing to the top of one wall in the room. Keep the North Pole pointing to the top of that wall the whole time as your Earth orbits the Sun.

Angle of tilt—23.5°   South Pole

4 You should be able to see a clear line between light and shadow/day and night on your Earth. It is vertical (straight up), not in line with Earth's tilted axis. The North Pole will be in darkness. If you now rotate your Earth on its axis to show day turning to night, you will see that the North Pole never comes into the daylight. The Arctic stays in darkness for weeks on end.

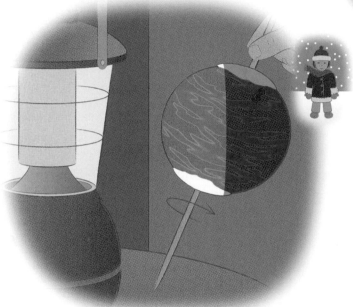

5 Stick the thumbtack in your Earth or stick a blob of sticky tack to the globe, somewhere between the Equator (the widest part of Earth) and the North Pole, then rotate Earth again. You'll notice that the pin/blob is in darkness more than light. This is midwinter in the northern hemisphere and northern countries have long nights and short cold days. It's colder because the Sun's rays are spread out at an angle over a wide area of Earth.

The opposite is true at the South Pole. When you turn the globe, it never gets dark there. Southern countries have long days and short nights, and it's warmer because the Sun shines down more directly on a smaller area. This is midsummer for the southern hemisphere.

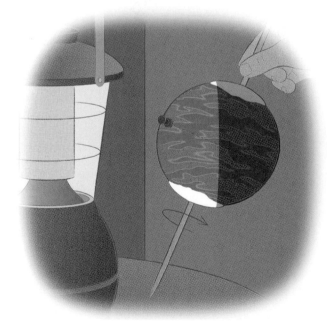

Stick the thumbtack on the Equator and you will see that day and night are the same length and the Sun is right overhead, making it hot.

6 Now, keeping the North Pole pointing at the same top corner of the room, begin to orbit your Earth in a counterclockwise direction around the Sun. When it is a quarter of the way round, stop. Here you can see that the line between night and day lines up with Earth's axis. Everywhere gets equal night and day. This is the equinox (which means equal night). The season has changed and it's warming up everywhere in the northern hemisphere—there it is spring. Everywhere in the southern hemisphere it is cooling down into autumn. At the Equator everything stays the same.

7 Move off on your orbit (with the North Pole still pointing at the same corner) for another quarter turn and stop again. Now you have the opposite of what you started with. The North Pole faces the Sun and has days when the Sun never sets. It is midsummer in the northern hemisphere and the days are long and warm and the nights are short. The South Pole is now in darkness. It is midwinter in the southern hemisphere. At the Equator everything is still the same.

8 Move on another quarter turn and you will find the light and dark line once again lines up with Earth's axis. This is another equinox: autumn for the northern hemisphere, spring for the southern hemisphere. Again, things stay the same at the Equator where days and nights are always 12 hours each and it stays hot all year round.

9 Now try this activity without the tilt. No seasons! What about with the axis on its side like Uranus?

Elliptical Orbits

You will need

An adult to help

Small plastic bottle with a lid

Sharp scissors with pointy ends

String

Broom handle and two similar chairs or somewhere to hang the pendulum

One or more big pieces of black or dark-colored paper

Packet of table salt or silver sand

Mixing bowl

Wallpaper paste (optional)

It is difficult to draw a proper ellipse, which looks like the path of a planet, with a pencil but it is easy and fun to draw ellipses using salt or sand and a pendulum. This way you can make brilliant orbit patterns. The longer the pendulum, the slower it moves—try it and find out!

1 Ask an adult to help you prepare the bottle. First, carefully cut off the bottom. Make a handle so the bottle will hang top down. To do this, pierce small holes on either side, thread some string through, and knot both ends to stop it pulling through. Make a small hole in the middle of the lid. It must be big enough for the salt or sand to run through freely.

2 Find a place to hang your pendulum. It needs to be somewhere high up where the pendulum can swing freely backward and forward and to the sides. A nail in the top of a door frame, a branch on a tree, or a bar on a climbing frame make for really long, slow pendulums. If you can't find anywhere, use two similar chairs with high backs. Place them back-to-back with the handle of the broom laid across their backs.

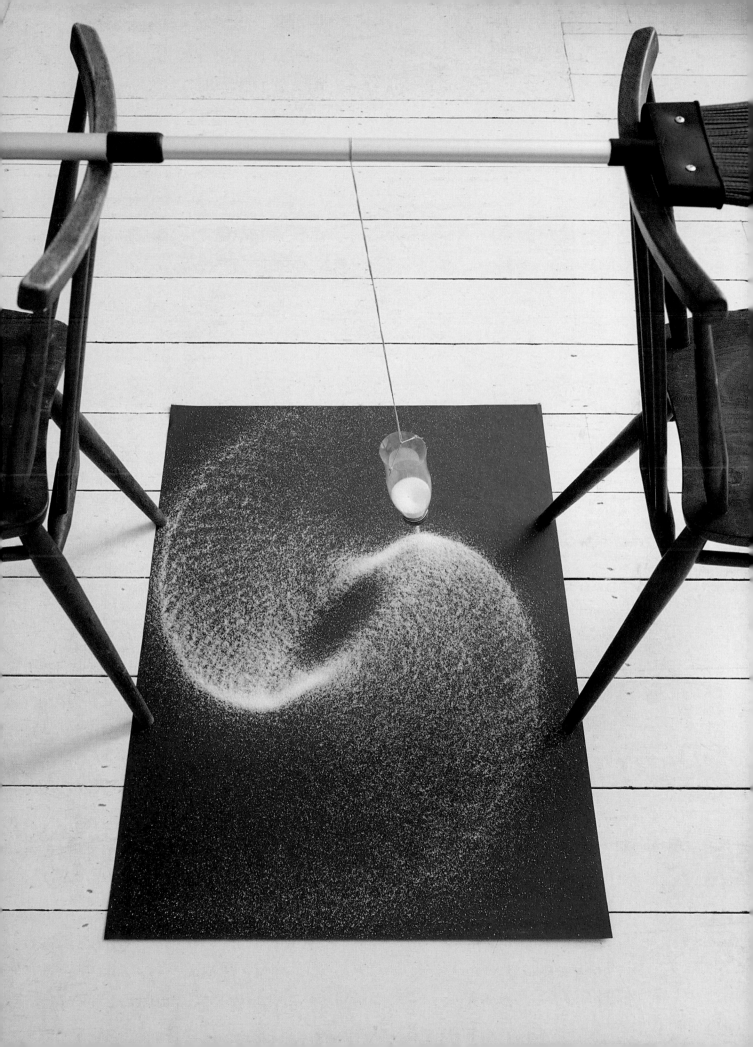

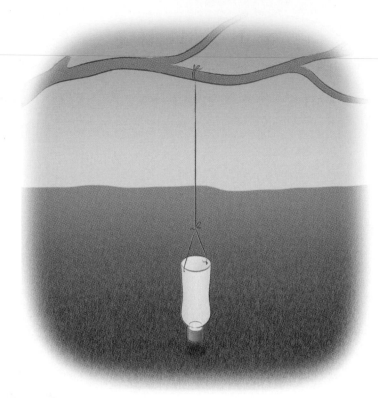

3 Ask an adult to help you to tie on the pendulum if your hanging place is high up. Tie one end of a long piece of string to your hanging point and the other to the middle of the bottle handle. Adjust the length of the string until the bottle hangs straight, about ½ inch (1cm) above the ground or floor.

4 Put a piece of dark paper on the ground under the pendulum bottle. The bottle should hang over the middle of it.

5 Practice swinging the pendulum bottle, so it swings in an ellipse over the paper, not just backward and forward in a straight line. Now hold your finger over the hole in the lid and pour salt or sand into the bottle until it is about 2 inches (5cm) deep. Take your finger off, swing the bottle again, and watch it draw beautiful ellipse patterns on the paper.

6 When the salt or sand run out, carefully lift the sides of the paper to collect the salt in the center and tip it into the bowl to use again. If you want to make a permanent pattern, spread wallpaper paste all over the paper before you set the pendulum swinging. Some of the salt or sand will stick to it. Catch the pendulum and cover the hole when you think your pattern is just right. Let the pattern dry and then shake the excess salt or sand into the bowl.

Chapter 4

The Moon

- -

THE QUESTIONS THIS CHAPTER ANSWERS:
Why does the Moon change shape?
Why are there shadows on the Moon's surface?
Why can't we always see the Moon?

- -

What is a moon?

Nowadays our streetlights are too bright for us to see the stars clearly, but we still notice the Moon. We are amazed at how big and bright a full moon can be and we love to see the thin sliver of a new moon, which looks like a fingernail clipping in the sky. Often, we can see the Moon even in daylight.

Our Moon is so big and so close that, with binoculars, we can easily see that the shadows on its surface are mountains, plains, and craters. It is also close enough to visit from Earth (only four days space flight away). Twelve astronauts walked on the Moon back in the 1960s and 1970s. Plenty more unmanned spacecraft have orbited or landed on the Moon since then and there are plans for more manned visits.

Our Moon is not the only one in the Solar System. There are hundreds of them. We call something a moon if it orbits a planet (or even a large asteroid) that is already orbiting the Sun. Mars has two moons, Jupiter has 79, Saturn 82, Uranus 27, and Neptune 14. There are also rocks and pieces of ice circling planets that are too small to be called moons. Saturn's ring system is made up of these. The bigger pieces are called moonlets; the rest are just called ring particles. Jupiter, Uranus, and Neptune also have rings made up of ring particles, but they are not so clear as Saturn's.

Moons came to orbit planets in different ways. Astronomers believe that Earth's Moon is made from pieces left over when a small planet, the size of Mars, smashed into Earth about 4½ billion years ago. The crash tipped Earth's axis into the tilt it has now and destroyed the small planet. All the pieces left over after the collision gradually gathered together to make our Moon.

The two moons of Mars were probably made in a similar way to Earth's Moon, by a big collision, but could have been asteroids pulled into its orbit by gravity. Jupiter's many moons are like a mini solar system. They are all very different from each other. The biggest moon in the Solar System is Jupiter's moon, Ganymede. It is bigger than Mercury and only a bit smaller than Mars. It would be a planet if it was orbiting the Sun. Earth's Moon is the next biggest.

Our Moon is so big and close, we can see the mountains, plains, and craters on its surface.

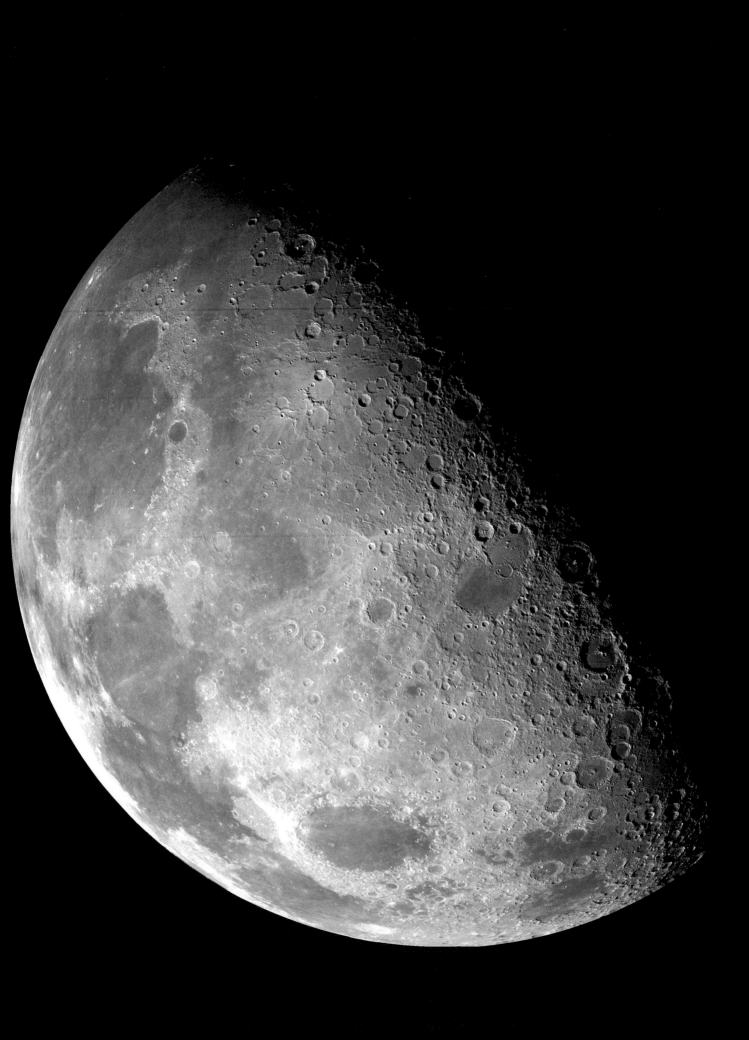

The phases of the Moon

The Moon has no light of its own. We see it because it reflects sunlight. Half of the Moon is always lit up by the Sun but because the Moon is orbiting Earth, we can't always see all the lit-up part from Earth. When the Moon is full it is on the opposite side of Earth to the Sun and all of it is reflecting light back to Earth. The Moon's orbit is slightly tilted away from the orbit of Earth around the Sun, so Earth doesn't usually block the sunlight. As it orbits Earth, it moves to the side and the part we see grows smaller and smaller until the Moon disappears. The Sun is now behind the Moon and we don't get any reflected light. The Moon continues on its orbit and starts to grow again until it is full. These changes in shape are called the phases of the Moon.

There are 29½ days from one full moon to the next. In history, this time became known as a month. Unfortunately, the 365¼ days of a year do not divide up exactly into 29½-day months, so the number of days in a month have been adjusted so that 12 months make a year.

The way the Moon rotates on its axis means that we only ever see one side, but we know what the other side looks like because spacecraft have orbited around it and taken photos.

The far side of the Moon

We only ever see one side of the Moon. We only know what it is like on the other side because spacecraft have flown around it and taken photos. The biggest crater is called the South Pole–Aitken Basin. It measures over 1,000 kilometers (621 miles) across and is on the side we can't see. We only see one side because the Moon rotates once on its axis in the time it takes to orbit Earth once. You can see how this works with a ball of modeling clay on a stick. Mark one side with a hole. Choose an object, such as a mug, to be Earth and take your Moon on an orbit around it. If you keep the hole pointing toward Earth, you will find that you have to twist the stick between your finger and thumb and that rotates the Moon on its axis. You need to rotate it once completely as the Moon orbits Earth.

Eclipses

The Moon is responsible for eclipses. In the past, people were very scared of these because they didn't know why they happened, but they are not difficult to understand. A solar eclipse happens during daylight. It is when the Moon's orbit takes it exactly in front of the Sun, so that when we look up from Earth the Moon blocks out the Sun and everything goes dark. These don't happen often. Sometimes, there is a partial solar eclipse when the Moon only covers part of the Sun.

Solar eclipses are so dramatic because of a strange coincidence. The size of the Moon and its distance from Earth means that it looks as if it covers the Sun exactly. If the Moon was closer to Earth, it would look much bigger, and you wouldn't see the edges of the Sun flaring out from behind the Moon. If it was farther away, it would look smaller and wouldn't cover the Sun. Total solar eclipses happen because the Sun is 400 times wider than the Moon but also 400 times farther away!

A solar eclipse happens when the Moon's orbit takes it exactly in front of the Sun and blocks its light. You must wear special glasses to watch a solar eclipse.

A lunar eclipse happens at night when the orbit of the Moon means that Earth comes exactly between the Sun and the full moon and prevents most of the Sun's rays from reaching the Moon. At the start of a lunar eclipse, we can see the shadow of Earth begin to cross the Moon. When the eclipse is total the Moon becomes an eerie red color because some light gets bent around, through Earth's atmosphere, toward the Moon (see page 93). Lunar eclipses happen more often than solar eclipses.

You can't see a solar eclipse everywhere on Earth at the same time. You have to be in the Moon's shadow and that isn't very big. You can find out the dates of the next eclipses where you live at https://www.timeanddate.com/eclipse/eclipse-information.html

The Phases of the Moon

The best (and tastiest) way to learn the phases of the Moon is to make an Oreo diagram—you can eat any broken cookies as you make it and the rest when you have finished with it!

You will need

Packet of Oreos

Sheet of paper

Small, blunt knife or popsicle stick, for scraping

Pencil

Notebook (optional)

1 Pull apart one Oreo very carefully so that you have a full circle of white frosting on a dark background on one side. That is your full Moon. The other side, with no frosting, shows the time when you can't see the Moon because the Sun is behind it—that's called the new moon. Put them both on a clean sheet of paper with the new moon on the right and the full moon opposite it on the left.

2 Now pull apart another Oreo. Use the diagram in the book (see page 50) to make a waxing crescent moon, carefully scraping away the frosting you don't need. Waxing is an old word that means growing. A crescent is a fingernail-clipping shape. When the Moon is waxing the crescent gets wider every day.

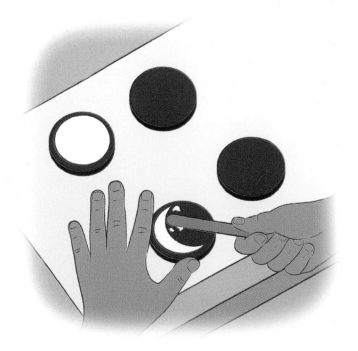

3 The next one to make is the first quarter. For this, cut the Moon circle in half down the middle and scrape half away.

4 Then make a waxing gibbous Moon. Gibbous is like a full moon with a crescent carved off the side. As it waxes, it keeps growing bigger and the piece carved off will get narrower each day until you reach the full Moon.

5 After the full Moon, the Moon begins to wane, which means it gets smaller. You have the same shapes as before but on the opposite side. Make a waning gibbous, a third quarter moon, and waning crescent and then you are back to a new Moon. Arrange them on your paper, as in the diagram, and label them all.

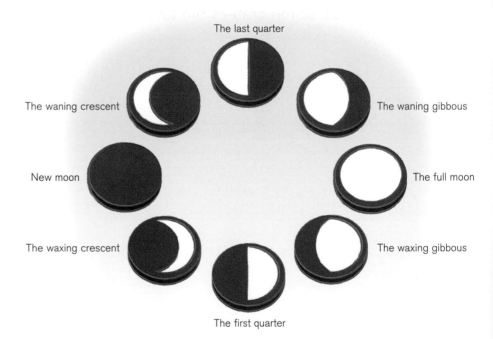

The last quarter

The waning crescent

The waning gibbous

New moon

The full moon

The waxing crescent

The waxing gibbous

The first quarter

6 If you are keeping a Moon diary, you can now draw the Moon each day that you can see it and label and date it correctly.

Crater Experiments

Using just binoculars, it is possible to see craters on the Moon and also patterns of rock around them (called ejecta patterns), which look a bit like rays of the Sun. There are millions of craters on the Moon. Some are tiny but there are at least a million that are more, often much more, than 1 kilometer ($^2/_3$ mile) across. They were created by meteorites crashing into the rocky surface of the Moon. Some were made billions of years ago. They haven't changed much since they were made because the Moon has no water or wind to wear them away or volcanoes to disturb them. The only change comes when a meteorite makes a new crater on top of an old one or tiny meteors knock into them. Not all are old, though. A new one has appeared at some point in the last 38 years between the times two spacecraft photographed the same place on the far side of the Moon.

But what makes some craters big and some small? Why do some have different ejecta patterns? Why are some deep and others shallow? Now is your chance to do some investigating and find out.

You can do this activity as a proper scientific investigation, carefully noting measurements and coming to some conclusions, or just have fun observing what happens.

You will need

White flour

Large tray, at least 1½ inches (4cm) deep (such as a large roasting pan)

Cake-decorating sprinkles or turbinado (demerara) raw cane sugar

Unsweetened chocolate (cocoa) powder or drinking chocolate powder

Strainer (sieve)

Old newspaper

Stiff tape measure

Marbles, ball bearings, or balls of modeling clay in different sizes

Paper and pencil (optional)

Cell phone (mobile phone) or digital camera (optional)

1 Pour some flour into the tray and level it out. It needs to be about 1 inch (2.5cm) deep.

2 Sprinkle the sprinkles or sugar in a thin layer all over the surface. Use a strainer to sift a thin layer of chocolate powder over the sprinkles. The different colors and size of grains will make the ejecta patterns easy to see.

3 Put a layer of newspaper on the floor (to catch spills) and place the tray on top. Hold the tape measure with its end on the surface of the flour. Drop a marble from a height of 20 inches (50cm) onto the flour. Look at the crater. How wide is it? How deep is it? If you are going to be scientific about this, measure it and note down the measurements. Look at the pattern of the ejecta (the flour and sprinkles) thrown up by the impact—this is what happens to the layers of rock on the Moon. You could take a photo to record the pattern.

4 Use the same marble, but this time drop it from a height of 3 feet (1m), into a different place. Is there a difference in the size of crater (measure it) or ejecta patterns? Try it from even higher. How does the depth and width of the crater change the higher the place you drop the ball from?

5 If there is still space on your tray, try dropping different sized balls all from the same height. Does the size of the ball make a difference to the size of the crater?

You could also try throwing balls from different angles or use small stones that are different shapes. Can you find/make small balls that are the same size but different weights to investigate?

6 After a while, once your whole tray is covered in craters, you could mix everything together to make pale brown flour (add a bit more chocolate powder to make it darker). Then sprinkle it with some different colored sprinkles with a new white layer of flour on the top. You can then do more investigating. Alternatively, you could try making craters on different surfaces. Fill the tray with silver sand instead of flour or try with wet sand or fine gravel. Use your imagination to keep investigating.

Chapter 5
Inside Planets

THE QUESTIONS THIS CHAPTER ANSWERS:
What is beneath your feet if you keep going down?
How does a compass work?
What is a shooting star?

Inside Earth

People used to think that Earth was flat and if you sailed over the horizon, you would drop off the edge. Now, of course, we know Earth is a sphere (ball). We have also discovered that what is on the inside of a planet makes a difference to what happens on the outside. Scientists can't yet travel to the center of Earth, or any other planet, but by studying the way planets behave, they can work out what is inside. They have worked out that, although Earth is a rocky planet with a solid surface, it is not all solid inside. There are four layers, which are all very different.

At the center is the inner core, which is a solid sphere made mostly of two metals, iron and nickel. It is as hot as the Sun's surface but the metal doesn't melt because the pressure inside Earth is so high it is squeezed together into a solid.

The outer core surrounds the inner core and is also made of very hot nickel and iron but, because the pressure is less here, it stays liquid. Around the outer core is the thickest layer, called the mantle. This is made of extremely hot rock, which is soft but not melted.

On the outside of Earth is the crust, which is very thin like the skin of an apple. Seventy percent of this is covered by the oceans. The crust is made of solid rock and is only about 30 kilometers (19 miles) thick on land. It is much thinner at the bottom of the oceans—about 5 kilometers (3 miles) thick. That is really thin compared to the whole Earth, which is 12,800 kilometers (8,000 miles) in diameter.

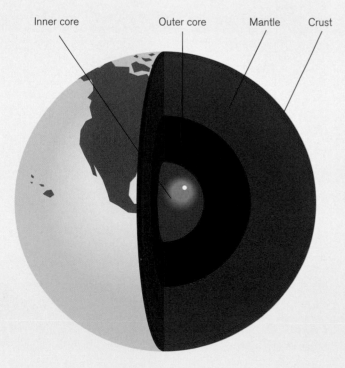

Inner core   Outer core   Mantle   Crust

Earth is a huge ball of incredibly hot rock with metal at the center. We live on a very thin skin of solid rock and water at the surface.

Volcanoes and earthquakes

The crust and the upper part of Earth's mantle are divided into 17 huge pieces, like a jigsaw, with some pieces much bigger than others. The pieces (called tectonic plates) move slowly around Earth at about the same speed as a fingernail grows. Where two plates meet there is a weak spot in the crust where pockets of hot magma (molten rock) which form in the mantle can push through to make volcanoes. When two plates grind past one another, ride up over one another, push together, or split apart, then we get earthquakes. Over billions of years, moving plates can push up whole mountain ranges like the Himalayas or move apart to create huge valleys like the Great Rift Valley in Africa.

Earth is not the only place with volcanoes. Venus has now been shown to have many

Volcanoes erupt where there are weak spots in Earth's crust and molten rock can push through from the mantle.

active volcanoes and Io, one of Jupiter's moons, has the most volcanic eruptions in the Solar System. No volcanic eruptions have been seen on Mars yet, but there were plenty in the past. Olympus Mons on Mars is the largest volcano discovered in the Solar System so far. Volcanoes have also been spotted on Triton, a moon of Neptune, and Enceladus, a moon of Saturn, but these throw out ice, not magma.

Volcanoes exist on rocky planets and moons, so why don't Mercury and our Moon have them? It seems they had volcanoes in the past but, as they cooled, their crusts shrank and the tectonic plates joined together to form one solid piece with no weak spots for magma to push through, meaning no more volcanoes.

Magnetic fields

The inner and outer cores are very important for life on Earth in a strange way. Because of heat from the inner core and the spin of Earth, the liquid metal in the outer core circles around in currents and this makes it magnetic. It is as if there is a huge, super-strong bar magnet threaded through Earth from pole to pole (but not quite in line with the axis). In the same way that two magnets repel each other, so Earth's magnetic field pushes away charged particles, which fly out from the Sun in what's called the solar wind. These particles would strip away an important part of our atmosphere called the ozone layer, which protects us from harmful rays of the Sun (see page 101).

The magnetic field is what gives us the auroras—the northern and southern lights. At the poles the magnetic field doesn't keep off the solar wind and the solar particles hit Earth's atmosphere, creating beautiful light displays in the night sky.

When the solar wind hits Earth's atmosphere at the North and South Poles, it creates the auroras, or the northern and southern lights.

Inside other planets

The other rocky planets Mercury, Venus, and Mars are made up of similar layers as Earth, but each is different. Mercury has a bigger core and thinner mantle; Venus has a very thick crust; and Mars does not have a liquid outer core.

The gas giants Jupiter and Saturn are not all gas. They have small cores of rock and ice but are mostly made of two gases, hydrogen and helium. The hydrogen in their mantles is under such pressure (squeezing) it has turned to liquid metal and this gives both these planets magnetic fields. Further from the core, the liquid hydrogen begins to turn to gas and you can't really tell where liquid ends and gas begins.

Uranus and Neptune, the ice giants, also have rocky cores, but have icy mantles and atmospheres made of hydrogen and helium.

On Jupiter, Saturn, and Neptune gases swirl around and create huge storms. Jupiter's Great Red Spot is an enormous storm which has been raging for at least 180 years. It is so big you could fit two Earths inside it! Uranus doesn't seem to be as stormy. This may be because it was in a massive collision billions of years ago which knocked it onto its side and that changed both how it is made up and how it behaves.

Asteroids and comets

Along with planets, asteroids and comets also orbit the Sun. Asteroids are lumps of rock and metal but comets are like huge dirty snowballs, several kilometers in diameter, made up of frozen gases, ice, rock, and dust. They mostly come from an area of space way out beyond Pluto (more than 2,000 AU from the Sun—see page 33). We only see one when it has been knocked into an orbit that takes it closer to the Sun. There, the Sun's heat begins to melt it and it develops a long tail of dust and gases, which streams out for up to 150 million kilometers (93 million miles). Comets can have such huge orbits that there may be hundreds of years between the times they can be seen from Earth. Halley's Comet is a famous comet that can be seen every 75 years.

Small parts from a comet's tail or small rocks from asteroids sometimes enter Earth's atmosphere. There they burn up and that's what makes a shooting star or meteor. If Earth passes through a part of a comet's orbit, then all the pieces of dust and rock it left behind can cause a meteor shower.

Some meteors are big enough to reach Earth. These are called meteorites. Scientists love to find these because some of them are leftover pieces from the beginning of our Solar System and can give us lots of interesting information about how it was formed 4.6 billion years ago. Others have been blasted off other planets by huge collisions. Several hundred from Mars have been found!

Jupiter's Great Red Spot—a storm twice as big as Earth.

A close-up of a comet—a huge dirty snowball made up of frozen gases, ice, rock, and dust.

Earth's Magnetic Field

This is a very simple activity but the science behind it is amazing!

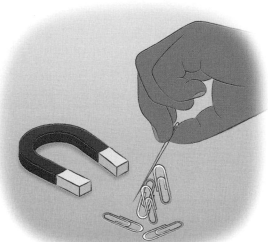

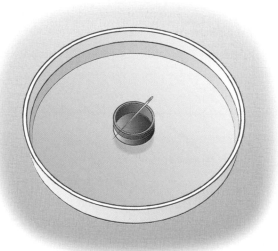

1 Stroke the needle with the magnet, always stroking in one direction, from the eye of the needle to the point, and lifting the magnet high above the needle as you circle it back. Stroke the needle at least 20 times. Now try to pick up some pins or paperclips with the needle to test if it has become magnetic. If it's not very strong, do some more stroking.

2 When the needle is magnetized, float the bottle lid like a boat on the surface of the water in the dish. Then carefully place the needle across the middle of the lid.

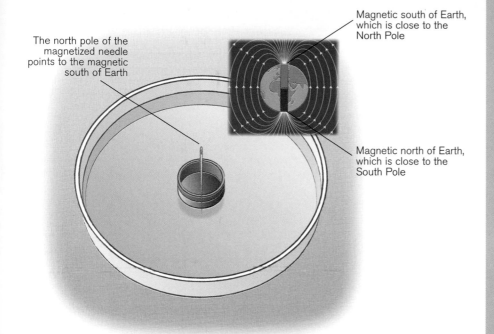

The north pole of the magnetized needle points to the magnetic south of Earth

Magnetic south of Earth, which is close to the North Pole

Magnetic north of Earth, which is close to the South Pole

You will need

Fine sewing needle

Strong magnet

A few pins or paperclips

Small plastic bottle lid

Wide shallow dish full of water

Compass or smart phone

3 Watch the lid as it spins around. It should come to rest pointing in a north–south direction. You won't be able to tell which is north unless you know which is the north pole of your magnet. The north pole of your magnet points to the magnetic south of Earth, which is close to the North Pole— it's a bit complicated!

4 You can check if the magnet is correct by using the Sun. If you live in the northern hemisphere, at midday the Sun is in the south and shadows point north. In the southern hemisphere, at midday, the Sun is in the north and the shadows point south. (You can also check with a store-bought compass or an app on a smart phone!)

Now think how amazing this is. Your compass works because the magnetized needle is lining up with the magnetic field produced by the liquid outer core of Earth 32,000 kilometers (2,000 miles) beneath your feet! But your magnet is not pointing exactly to true north because that is slightly different from magnetic north. You have to find that by using the stars (see page 110).

Volcanoes on Venus

Whether a planet has volcanoes can tell us a lot about what is inside it. In this activity, you will make a Venusian volcanic landscape, which includes some pancake volcanoes that are found nowhere but Venus.

1 First make a cone-shaped volcano like we have on Earth. Stand the bottle on the piece of cardboard but not right in the center. Scrunch up newspaper into balls and place these around the bottle. Tear off strips of masking tape. Stick one end of each piece of tape just inside the neck of the bottle and stretch the other end out at an angle over the newspaper to the edge. Do this all around the bottle to make the cone shape, adding more paper balls where they are needed.

2 Mix some PVA glue with water to thin it down. Tear newspaper into strips and paste these all over the volcano. Build up two or three layers of paper to make the volcano strong. Make sure you haven't filled in the hole at the top of the bottle—you need the top to be clear for the eruption.

You will need

Small plastic bottle

Large piece of cardboard cut from a cardboard box, as a base

Newspaper

Masking tape

Mixing bowl

Paintbrushes

PVA glue

3 or 4 small, paper party bowls (or yogurt pots with the sides cut shorter)

Acrylic paint set

Baking soda (bicarbonate of soda)

Pitcher (jug)

Vinegar

Dishwashing detergent

Red food coloring

Funnel (optional)

3 Now make some pancake dome volcanoes. These have flat tops and steep sides. Arrange the party bowls in a group and stick them down with strips of masking tape. Paste newspaper strips all over these until you have a whole papier-mâché landscape. Let dry.

4 Now paint the landscape in shades of brown, yellow, and gray. Paint lava flowing out from the top of the cone volcano and down the sides in bright red and orange. Lava does not spill out of pancake dome volcanoes—it just builds up underneath into the dome shape, so don't paint lava on these. Let the paint dry. Finally, varnish the whole landscape with a coat of undiluted PVA glue to waterproof it. Let dry.

5 Now for the eruption. Drop 2 tablespoons of baking soda into the bottle. Mix half a cup (120ml) of vinegar with a squirt of dishwashing detergent and a few drops of red food coloring in the pitcher. Pour the vinegar mixture quickly into the bottle. Use a funnel if you have one but take it out quickly. Stand back and watch the lava flow out.

Meteors and Friction

Friction is the force you get when one surface slides over another. There is a lot of friction between your rubber-soled shoes and the ground, so you don't slip over. There is very little friction between skis and snow, which is why you can slide so easily. Where there is friction, you get heat. You can feel this just by rubbing your hands together or touching the brake blocks on your bike after you have braked on a hill (don't burn your finger!).

When asteroids, comets, and spacecraft that are returning to Earth enter Earth's atmosphere, the air slides over their surfaces. Because they are going so fast (40,000–260,000 kilometers per hour/25,000–160,000 miles per hour) there is a lot of friction and a lot of heat, enough to burn them up—that's why we see them as shooting stars. Spacecraft returning to Earth need to have heat shields to stop them burning up when they re-enter Earth's atmosphere.

In the past, people used the power of friction to build fires. Have a go. It's a useful survival skill but it's not easy! This activity will take you halfway to starting a fire. Be very careful.

You will need

A long, dry branch, around 2-3 feet (60-100cm)

Sharp stone (or a penknife if an adult is helping you)

Strong, dry stick

1 Peel the bark off one side of the branch and, always pushing the stone away from you, use it to shave off some of the wood underneath to make a flat surface.

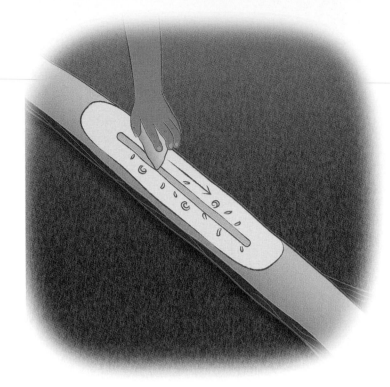

2 Use a point of the stone or a different stone and, still pushing away from you, cut a groove into this flat area. Make it about 8 inches (20cm) long.

3 Peel the bark off the smaller stick and sharpen one end of it into a point. Keep cutting away from you!

4 Sit on one end of the branch and push the other end into soft ground to hold it steady. Then, with both hands on the stick, one on top of the other, hold the pointed tip of the stick in the end of the groove and push it quickly backward and forward. Keep the stick angled back toward you, not upright. You don't need to push hard, just move quickly. The heat will begin to increase, and a small pile of wood shavings will build up at the end of the groove. Keep pushing harder and faster and you should begin to get smoke!

Find the dates of meteor showers on https://earthsky.org/astronomy-essentials/earthsks-meteor-shower-guide.

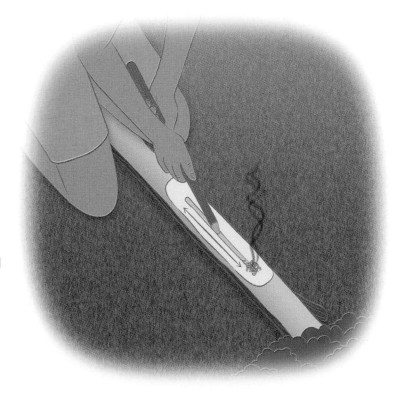

Chapter 6
Atmosphere

- -

THE QUESTIONS THIS CHAPTER ANSWERS:
What is above your head if you keep going up?
Why do we get wind and clouds?

- -

The air we breathe

When we think of air we usually think about breathing, but the atmosphere does much more for Earth than simply give us air to breathe. In fact, if we time-traveled back more than 500 million years, we would need respirators because there would be no oxygen. All the oxygen in the air has been made by plants in a process called photosynthesis (see page 115).

Weather

Weather is an important part of our atmosphere. The Sun's heat warms the surface of Earth and then the air above it gets warm. Warm air rises and then cold air rushes in to take its place and we get winds.

The same happens in the oceans. Warm water rises, cold water sinks, and we get currents that circle Earth. These currents affect the weather all over the world, warming it or cooling it.

Heat from the Sun on the oceans makes water evaporate and change into water vapor, which mixes with the air as a gas. Higher up in the atmosphere the air cools and the water vapor changes back to tiny water droplets, which we see as clouds. These are moved across the sky by winds in ever-changing patterns. When the water droplets get too big and heavy, they fall as rain, snow, or hail and give Earth the fresh water needed by plants and animals that live on land.

The six layers of Earth's atmosphere

The atmosphere also acts like a quilt and keeps us warm. It is made up of six layers surrounding the whole Earth.

To understand it, you need to know a bit about molecules. Break up anything you have into smaller and smaller and smaller pieces and eventually you will break it into atoms, which are a kind of chemical building block. Atoms combine with similar or different atoms to make molecules. In a solid, these molecules are packed tightly together and don't move about but, in a gas, they are widely spaced and ping about all over the place. Because they are widely spaced you can squash them together. That's what happens when you pump air into a bike tire: you squash a lot of air molecules into a small space and that creates very high pressure. If you open the valve, the air rushes out and tries to spread out again.

Clouds of water vapor swirl through the inner layer of Earth's atmosphere.

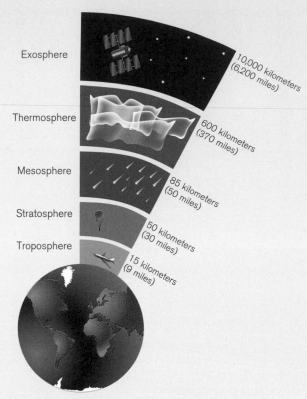

Exosphere

10,000 kilometers
(6,200 miles)

Thermosphere

600 kilometers
(370 miles)

Mesosphere

85 kilometers
(50 miles)

Stratosphere

50 kilometers
(30 miles)

Troposphere

15 kilometers
(9 miles)

We live and breathe in the first layer of the atmosphere called the **troposphere**. It stretches for up to 14 kilometers (9 miles) above our heads. It is much, much thinner than any of the other layers. It is mostly made of molecules of the gas nitrogen, but with about one-fifth oxygen and small amounts of carbon dioxide, water vapor, and a gas called argon.

Where we are, in the troposphere, there are a lot of gas molecules which are fairly close together. Of all the molecules in Earth's atmosphere about three-quarters of them are in the troposphere. As you travel up through the atmosphere there are fewer gas molecules spaced farther apart. That's why it is difficult to breathe at the top of high mountains.

Above the troposphere is the **stratosphere**. This is the layer of ozone gas that protects us from the Sun's harmful ultraviolet rays (see page 101). This is another reason why Earth's atmosphere is so important. Most big jets fly at the very top of the troposphere or the bottom of the stratosphere because there are no clouds and no changing weather there. Also,

because there are fewer molecules to push through, the planes use less fuel.

A shooting star is a meteor traveling through the **mesosphere**, which is the next layer up. The top two layers of the atmosphere contain so few molecules that there isn't much friction (see page 65) and the meteor gets through without burning up. Only when it hits the mesosphere does it start to glow.

We think of the International Space Station as being in outer space, but it is actually still in Earth's atmosphere in the **thermosphere**. The atmosphere here is very thin and there is no oxygen to breathe. A lot of satellites are also orbiting in the thermosphere. These are used for things such as communication, weather forecasting, and collecting data about Earth. They are eyes in the sky!

The last layer is the **exosphere**, which is enormous and contains gases like hydrogen and helium, but the molecules are so widely spaced that they never bounce into each other. As you move farther and farther away from Earth there are fewer and fewer molecules until you reach the vacuum of space, but even here there are some gas and dust molecules.

The outer three layers of the atmosphere are also all part of the **ionosphere** and this is where the auroras light up the sky. Ions are molecules which become charged. Have you ever rubbed a balloon on your hair and then lifted it away to make your hair stand on end? Rubbing hair on a balloon creates charged molecules and, like magnets, these can repel or attract one another. The ionosphere is important because ions affect radio waves. They get bounced off them, bent by them, or absorbed by them. The ionosphere changes constantly, even between day and night, which can cause problems for satellite communication and GPS which use radio waves.

Greenhouse gases and global warming

Our atmosphere keeps us warm. The Sun shines through it and heats up the planet. The heat is radiated back to space (like from a radiator) but the atmosphere bounces some of it back to Earth.

In the last 200 years, our atmosphere has begun to change. Whenever we burn gas, coal, and oil for transport, factories, heating, or to make electricity, carbon dioxide gas is produced and becomes part of the atmosphere. Billions of farm animals belch out methane into the atmosphere. These gases are called greenhouse gases. That's because in a greenhouse the Sun shines through the glass and warms up the inside but the glass stops the heat from escaping. Greenhouse gas molecules are very good at stopping heat from the Earth escaping into space, so the Earth is getting warmer. Everyone is worried because a warmer Earth will mean more droughts, heatwaves, wild fires, storms, floods, and melting glaciers and ice caps.

In the 20th century, a gas used to make fridges was found to be destroying the ozone in the stratosphere. Politicians got together and agreed not to make this gas anymore and gradually the ozone layer is returning to what it was before. We found we could solve a problem. Global warming is a much bigger problem, however, and we must find ways to solve it.

Atmospheres on the other planets

Mercury should be the hottest planet in the Solar System because it is closest to the Sun, but Venus is hotter. That's because Mercury has no atmosphere to reflect the Sun's heat. Planets need strong gravity (see page 80) and a magnetic field (see page 56) to hold onto an atmosphere. Mercury has neither of these.

The Sun heats up the Earth, but the heat is radiated back to space. The atmosphere stops some of the heat from escaping and keeps us warm. Greenhouse gases in the atmosphere reflect too much heat back to Earth.

Venus shows us how a planet can heat up because of the greenhouse effect. Venus is quite close to the Sun but it also has a dense atmosphere made mostly of carbon dioxide (with clouds made of acid droplets), which is very good at reflecting heat. On the surface of Venus, it's hot enough to melt the metal lead.

Mars used to have an atmosphere but billions of years ago it lost its magnetic field, so most of its atmosphere was stripped away by the solar wind (see page 56). It is much colder than Earth because it is farther from the Sun and its very thin atmosphere (made mostly of carbon dioxide with a little oxygen) can't keep it warm.

The four outer planets have atmospheres that are made mostly of hydrogen and helium gases, with clouds made not from water but from drops of methane and ammonia (another gas), which have turned to liquid.

The Power of Air Pressure

We don't usually notice air except when the wind blows. We don't notice that air is pushing on us from all directions because we have never known anything else. The push of the air is called air pressure. These two quick activities prove that air pressure exists! The first you can do on your own. You'll need an adult to help you with the second one.

1 Put the beaker in the kitchen sink and fill it with water. Use the pitcher to keep slowly adding water until it seems to bulge up higher than the top of the beaker.

You will need

An adult to help you

Plastic beaker

Pitcher (jug)

Piece of card (slightly bigger than the top of the beaker)

Plastic bottle with a lid

Bowl of ice cubes

2 Carefully slide the piece of card over the top of the beaker. Hold the card on the beaker with one hand as you lift it and turn it over with the other. Keep the beaker over the sink! Now take your hand away from the cardboard. No water should fall from the beaker. Air pressure pushes from all directions. It pushes up more strongly than the weight of the water pushes down. If you let in the tiniest bit of air, all the water will fall out.

3 Now you must get an adult to help you. Ask them to pour an inch of boiling water into the plastic bottle, put on the lid (not too quickly), and place the bottle in the bowl of ice cubes. Don't touch the bottle at all but watch what happens. As it cools it begins to crumple in on itself. That's because the water vapor from the boiling water pushed out the air molecules. Then, as it cooled, the water vapor turned back to water. There are fewer gas molecules in the bottle, so the pressure is lower. The higher-pressure air outside the bottle pushes on the bottle and crumples it up.

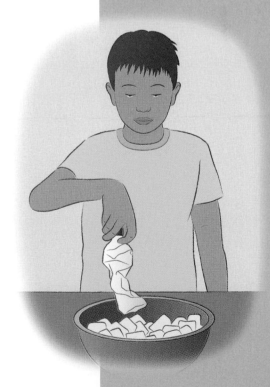

Circling Currents

A convection current is the name of the circling current you get when warm gas or liquid rises and cold gas or liquid falls. They rise because when they get warm, the molecules move faster and spread out more. That means there are fewer molecules in the same space, so the gas or liquid becomes lighter. When they cool down the molecules move closer together again, so the fluid gets heavier.

Convection currents are important for our world. They happen in Earth's liquid outer core and in its mantle and help to create our magnetic field (see page 56). They happen in the oceans and create ocean currents and they happen in the air to create wind and rain. These two activities show convection currents in water and air.

1 Mix some food coloring into a little water so the water is very dark and pour it into a few of the ice cube tray holes. Fill some of the other holes with uncolored water. Freeze overnight.

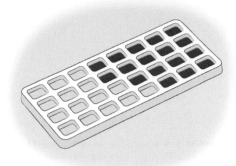

To show convection currents in water you will need

Dark food coloring

Ice cube tray

Large 2-quart (2-liter) clear plastic bottle (or a fish tank if you have one)

Sharp scissors

Spice jar with holes in the lid or a small pepper shaker

2 Ask an adult to help you cut the top off the plastic bottle and fill it with cold water.

3 Pour a little food coloring into the pepper shaker and fill it right to the top with hot water from the tap. You want it to be hot but be careful not to burn yourself.

4 Drop the pepper shaker into the bottle of water. You may have to tap it a few times to knock out any air bubbles. Color will stream up out of the pepper pot and swirl up toward the top of the bottle as the warm water rises. (Try it with cold water and nothing will happen.)

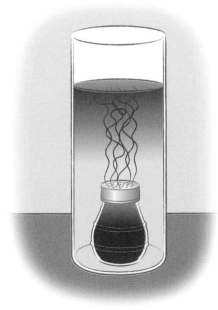

5 Drop some ice cubes into the water and you may see the color sink down again as the water cools. If the color has spread out, this may be hard to see, so now drop in the colored ice cubes and watch how cold water sinks.

To show convection currents in air you will need

Pair of compasses

Piece of thin colored card

Pencil

Bamboo skewer

Scissors

Note: If it is summer and there aren't any hot radiators, you will need

An adult to help you

Ball of modeling clay or sticky tack

Large metal baking tray

Tealight candle

1 Use a compass to draw a circle on the piece of thin card, about 3 inches (8cm) across. Mark the center where the point of the compass was.

2 Starting near, but not on the center point, draw a spiral on the circle. You need about three turns and then the line should reach the edge of the circle. Cut along the line.

3 Fold the inside of the spiral a little in a cross shape with the center in the middle of the cross, making a point. Thread the skewer up through the turns of the spiral and balance the point of the spiral on the top of the pointed end of the skewer.

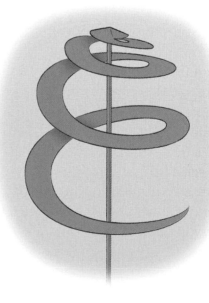

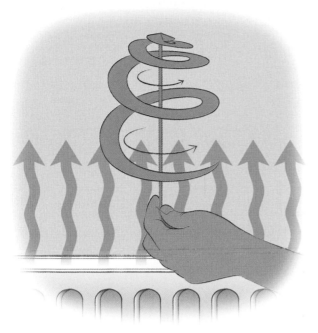

4 If you hold this above a radiator, you will see the spiral begins to spin because hot air is rising and pushing it round. Move the spiral away from the radiator and it will stop spinning.

5 Alternatively, if there are no hot radiators, push the end of the skewer into the ball of modeling clay to keep it upright. Place it onto the baking tray for safety. Put a tealight under the spiral and ask an adult to light it. The hot air rising from the candle will send it spinning. Be very careful that it doesn't drop into the candle and catch fire. Blow out the candle after you have watched it spin. Use a big square baking tray for greater safety.

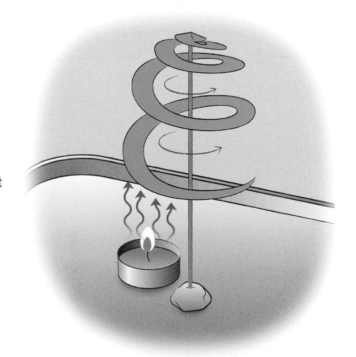

Making Methane

Methane is one of the main gases causing global warming, along with carbon dioxide. There is not as much methane going into the atmosphere, but it has a much stronger warming effect than carbon dioxide. You make methane every day! Methane is one of the gases that comes out of your bottom when you fart! Cows make much more than we do. Most comes out as burps but some as farts. Anything that rots also produces methane because bacteria are eating it and they produce methane as a waste product.

Methane is one of the gases found on Jupiter, Saturn, and Uranus. Neptune looks blue because of the methane in its atmosphere.

1 Ask an adult to help you. Collect kitchen scraps such as potato peel, apple cores, banana skins, the outside leaves of lettuces and cabbage, onion skins, and orange peel. Cut everything up small and put it in the blender with a cup of water. Do not overload the blender!

An adult to help you

Kitchen scraps from vegetables/fruit only (anything suitable for a compost bin)

Kitchen blender

Funnel

2-quart (2-liter) plastic bottle

Balloon

2 Blend everything into a "compost smoothie." Add more water if you need to. Use the funnel to pour it into the bottle. Half-fill the bottle.

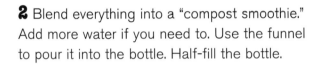

3 Stretch the top of the balloon over the neck of the bottle. Place the bottle somewhere warm and leave it for a few days. Keep checking it. Bacteria will get to work eating the vegetables/fruit and the balloon will gradually begin to inflate! It will be full of methane.

Chapter 7
Gravity

- - - - - - - - - - - - - -

THE QUESTIONS THIS CHAPTER ANSWERS:
Why do things fall when you drop them?
Why doesn't the Moon fall down to Earth?
Is it possible to see a black hole?

- - - - - - - - - - - - - -

What is gravity?

Nothing that you see around you would exist without gravity. It was gravity that first pulled together all the bits of rock and gas that make up the planets. It is gravity that keeps everything in the Solar System in orbit around the Sun and it is gravity that holds everything down on the surface of Earth, so it doesn't fly away into space.

Isaac Newton (see page 123) was the scientist who first realized the importance of gravity and how it worked. The story goes that he was sitting under a tree and an apple fell on his head. He began to understand that gravity is the force that pulls any two objects together, and that each object pulls on the other one. Gravity is a weak force, which means that for two small objects it can't be felt, but when you have an object as big as a planet or the Sun, the force becomes enormous. He also realized that gravity becomes weaker the greater the distance between the two objects, in exactly the same way that magnets pull each other less strongly when they are farther apart.

Another scientist called Albert Einstein (see page 123) worked out more about gravity in space, but you need to learn some difficult math before you can really understand his ideas. Scientists know all about how gravity works but are still puzzled about why it works!

Isaac Newton

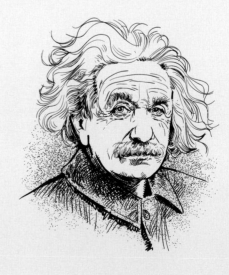

Albert Einstein

Gravity and tides

When you're at the beach, have you ever thought why tides come up and cover the sand twice a day? It is to do with gravity. Earth and the Moon are two enormous objects that are quite close to each other and so gravity pulls each way. When the Moon's gravity pulls on the ocean, it makes it bulge out on the side closest to the Moon and also on the opposite side. As Earth turns, the bulges keep in line with the Moon and so move around Earth and the tides go in and out. When the Sun and the Moon are both in line with Earth their gravity works together to make a bigger bulge and a higher tide called a spring tide. When they are not in line there are lower tides called neap tides.

Earth pulls back on the Moon, but there is no water on the Moon to make tides. Instead, the Moon itself is very slightly stretched toward Earth, and, as well as the oceans bulging, Earth gets a tiny bit stretched toward the Moon. Because of this, the closest part of the Moon has become locked in position so that it always faces Earth. That's why we never see the other side of the moon as it orbits around us (see page 46).

Gravity and weight

Because Earth is so big, everything on Earth is pulled strongly toward its center. That's why if you throw something up, it always comes down and, wherever you are on Earth, down is the center of Earth. But how much Earth pulls depends on the size of the object it is pulling on. It is hard work to lift a heavy object off the ground because Earth is pulling down on it so strongly. It's much easier to lift a light object because Earth is pulling on it less strongly. What we call weight is the size of the force that is pulling the object down toward the center of Earth. Although it is hard to imagine, everything on Earth, including you, is also pulling Earth toward it, but the force is very, very small.

If you were to travel to another planet your weight would be different. The stuff you are made of, which is called your mass, wouldn't have changed, but the planet would be pulling on you by a different amount. Gravity is stronger on planets with more mass. So, if you weigh

Astronauts could bounce around on the Moon because the Moon's gravity is only one sixth as strong as the Earth's.

66lb (30kg) on Earth, on Jupiter, the biggest planet in the Solar System, you would weigh 167lb (75.8kg) and it would be hard to lift up your feet (although, of course, you couldn't land on Jupiter because it has no surface). On Mercury, the smallest planet, you would only weigh 25lb (11.3kg). Some planets with much more mass can still have similar or even less gravity than Earth at their surface. They are so big that where you would measure their gravity is much farther from their center, and remember, gravity gets smaller with distance.

When astronauts landed on the Moon they could bounce around in that strange way because the Moon's gravity is only about one-sixth of Earth's. When they jumped up, they were only pulled down very gently.

Gravity and space travel

To get to the Moon, or anywhere in space, rockets have to escape Earth's gravity, which isn't easy. They need huge amounts of power to reach a speed of around 40,000 kilometers per hour (25,000 miles per hour)—that's about 120 times faster than the take-off speed of a jumbo jet—to break free of gravity, which is why a rocket taking off looks so spectacular and is scary for astronauts. Luckily, it is easier to leave the Moon to come back to Earth because its gravity is so much less.

Gravity also becomes weaker the farther two objects are from one another, but you need to go a long way before you feel much difference. When astronauts landed on the Moon there wasn't much left of Earth's gravity pulling on them, but on the International Space Station, which orbits only 400 kilometers (250 miles) above Earth, the astronauts are still in Earth's gravitational field. They feel weightless because the space station is always free-falling toward Earth. It is only because it is orbiting so fast (it circles Earth 16 times every day) that it never falls down to hit Earth. Instead, it orbits in a curve that roughly matches the curve of Earth's sphere. For exactly the same reason, the Moon never falls from the sky!

Living without gravity can be fun but it is difficult—sneeze and the drops you spray out would float around the spacecraft forever rather than falling to the ground! You can't pour out a glass of water and you have to tie yourself to something to sleep or you might float off. You have to keep exercising, or your muscles get weak from not having to use them to move your body around, and "How do you use the bathroom?" is the first question everyone asks an astronaut!

A rocket needs huge amounts of power to escape Earth's gravity.

Gravity and the Sun

The Sun is bigger and has more mass than anything else in the Solar System, and it doesn't have just a bit more mass. Imagine all the stuff that makes up the mass of the Solar System divided into one hundred equally sized balls of "playdough." Now pull a little piece off one of the balls and squash the rest of that ball together with the other 99 balls. To produce a model of the Solar System, you would have to make all the planets, dwarf planets, asteroids, and comets out of the little piece you pulled off and the rest of the playdough would become the Sun! That's why the Sun's gravity is able to keep everything in the Solar System orbiting around it. Planets don't fall into the Sun because they are orbiting so fast (in the same way that the Moon and the International Space Station don't fall to Earth).

Black holes

Stars do not last forever. Just as a fire burns out when it runs out of wood, a star comes to the end of its life when it eventually runs out of its fuel, hydrogen gas. What happens next depends on the size of the star, but the very biggest ones explode in a massive explosion called a supernova. When the dust clears, what may be left is a black hole. The huge mass of the star has been pulled into a very small space. This means that if you get too close to a black hole its gravity is unbelievably strong. Everything that gets too close, even light, is pulled into it and cannot escape. You can't see a black hole because it doesn't reflect any light. Astronomers know where black holes are because other stars behave as if there is something pulling them with huge amounts of gravity, but whatever is pulling them is invisible!

Asteroids and comets

Large asteroids and comets zooming through space on strange orbits sometimes pass close to Earth. Earth's gravity might pull them onto a slightly different course, but usually they are traveling so fast they are not pulled down to Earth. It does happen though! An enormous crater has been found in Mexico which was made by an asteroid hitting Earth 66 million years ago. The impact filled the atmosphere with dust that blocked out the Sun, so Earth's temperature dropped. This probably caused the extinction of the dinosaurs. Today astronomers keep a close eye on asteroids and comets that might come close to Earth, but none have been found that are likely to collide with us.

Astronomers were able to watch a comet called Shoemaker Levy flying into Jupiter in 1994. About 30 years before, Jupiter's gravity had captured it so that it began to orbit Jupiter rather than the Sun. Jupiter's gravity pulled on it so strongly that the comet broke apart into

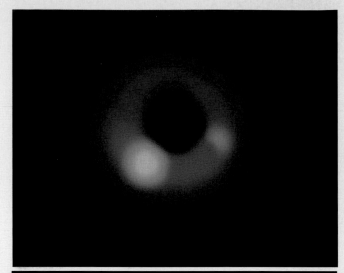

TOP: Even light cannot escape from a black hole.
BELOW: An image of comet ISON taken by the Hubble Space Telescope.

a long trail of pieces and two years later it plunged toward the center of Jupiter in a huge fireball.

Some astronomers think that we should be grateful to Jupiter. Because it is so big, they think Jupiter might pull comets and asteroids away from Earth and protect us from impacts. Nowadays, other astronomers have begun to think the opposite—Jupiter might just pull them into orbits closer to us!

OPPOSITE: Comet Shoemaker Levy on a collision course with Jupiter.

Fun with Gravity   Four Quick Experiments

It is not difficult to show gravity working. Just drop something, jump up, or trip over! Here are a few very quick activities to help you understand some more things about gravity.

Experiment 1
Gravity and air resistance

You will need

Big stone

Small pebble

Thin aluminum food tray (from a takeout)

1 Work with a friend. Drop the big stone and the small pebble from high up at exactly the same time. Do they hit the ground at the same time?

2 Drop the big stone and the aluminum tray at the same time. What happens to the tray? Do the same experiment with the tray upside down. What happens this time?

Behind the science
Many people think that heavy objects fall faster than light ones. They are wrong. Gravity makes everything fall at the same speed. If you are on the Moon and drop two different things at the same time, however heavy one is and however light the other, they will hit the ground at exactly the same time (an astronaut tried this with a hammer and a feather).

On Earth it's different because of our atmosphere. Pushing through air molecules slows things down (see pages 68–70). This is called air resistance and is why a parachute works. The foil tray has a lot of air resistance and you can see the air pushing it around as it falls. Scrumpling up the foil makes it smaller and reduces its air resistance, so it lands at the same time as the stone. Galileo (see page 123) was the first to realize this about gravity and is said to have proved it by dropping things such as cannon balls and musket balls from the top of the Leaning Tower of Pisa in Italy.

3 Scrumple up the tray as tightly as you can into a small ball. Drop the aluminum ball and the big stone at the same time. What happens now?

Experiment 2
Gravity and asteroids

You will need

Strong magnet

Shiny tray

Sticky tape

Books to prop up the tray

Steel nuts, washers, and paperclips in different sizes

1 Stick the magnet to the center of the tray with a piece of sticky tape and prop the tray up at one end so that it is like a slide.

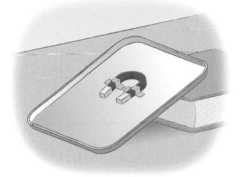

2 Slide the nuts, washers, and paperclips down the tray one at a time. Start at the side of the tray, well away from the magnet, then get closer and closer to it. When does the magnet start catching the metal objects? Does it catch lighter ones when they are farther away than heavy ones? Does the steepness of the tray make it harder for the magnet to catch them? Can you get the nuts to swerve around the magnet but not get caught?

Behind the science

Magnetism is not the same as gravity but it is a good model for an invisible force that pulls on objects. Remember, gravity pulls on everything, not just magnetic metals, but it is much too weak to have an effect on tiny washers.

This model shows how asteroids and comets can be pulled in toward a planet. The swerving nuts show how a planet's gravity can nudge them off course into a different orbit in the same way Jupiter is believed to affect asteroids and comets heading toward Earth.

Experiment 3
Gravity and orbits

You will need

Small beach bucket (or similar) with a handle

String

1 Tie about 24 inches (60cm) of string to the handle of the bucket.

2 Go outside to the center of your backyard and, with everyone else well out of the way, swing the bucket in a fast vertical circle so it is upside down at the top of the circle. When the bucket is at the top, let go of the string. What happens to it?

Behind the science

This is a model to help you understand how planets keep orbiting the Sun, but are not pulled into it, and how the Moon and the International Space Station orbit Earth and are not pulled down to the surface.

The water in the bucket and the bucket are pulled toward Earth by gravity. They should drop down when they reach the top of the circle, but they are moving very fast. They have what is called inertia. If left alone, they have to keep going in the direction they are moving which, at any point, is straight along the edge of the circle.

That's why when you let go of the string the empty bucket shot off across your backyard! When you are holding it, the string pulls the bucket back toward the center of the circle and stops it flying away. When there is water in the bucket, the water can't fly off because it is stopped by the side of the bucket. In this model, the string is like gravity holding planets and everything else in orbit, always pulling them toward the center to make them move in a curve.

3 Now half-fill the bucket with water. Again, swing the bucket in a fast vertical circle. This time don't let go! Even though the bucket is upside down at the top of the circle, you should not get wet!

Experiment 4
Space–time and gravity

You will need

Mixing bowl

Pair of old adult-sized pantyhose (tights)—ask first!

Large rubber band to fit tightly around the rim of the bowl

Scissors

Marbles or balls of modeling clay in different sizes

1 Push the bowl inside the pantyhose so the stretchy fabric is stretched tightly across the top of the bowl. Stretch the rubber band over the top of the bowl around the rim to hold the fabric in place.

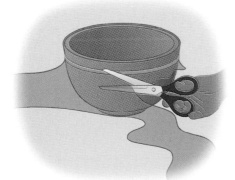

2 Cut off the remainder of the pantyhose so that the bowl sits flat on a table.

3 Put a large marble or ball of clay in the center of the bowl (this represents a big planet in space). It will make a dip in the fabric. Now put another marble on the bowl. What happens? Now experiment with rolling a small marble past the big one. Can you get it to orbit the big marble a few times rather than be pulled straight into it?

Behind the science

Albert Einstein (see page 123 said that gravity was a warp in the fabric of space–time! You have to start thinking differently to understand space–time. It is the three dimensions of space plus time, which makes a fourth dimension. It is usually shown in diagrams as looking and behaving a bit like the stretchy fabric of the pantyhose. Gravity works in a similar way to the dips the marbles make, which made them roll together. If one marble is moving fast, it may begin to orbit the other rather than just collide. Space–time is useful for explaining things like tiny movements in stars or for tracking the exact movement of satellites around Earth.

You will need

An adult to help

Rokit water rocket kit (available online)

Large, plastic bottle (1 or 2 quarts/liters)— check the Rokit collar fits onto the bottle

Bicycle pump (a floor pump is best)

Bucket of water and a pitcher (jug) for filling

Large open space for launching (check there are no overhead power lines)

Stopwatch or smart phone (optional)

Blast Off a Water Rocket

If you have ever inflated a balloon and then let it go before you have tied up the neck, you know it shoots off around the room. The escaping air blows out one way and creates thrust which pushes the balloon off in the other direction. Thrust is what blasts rockets into space. Exploding fuel pushes down and the rocket shoots up.

It takes a huge amount of force to shoot a rocket into space against gravity. You need quite a lot of force to get anything high into the air because gravity pulls it down. Like a balloon, this water rocket uses the power of air pressure to create thrust and blast-off, though not nearly fast enough to reach orbit.

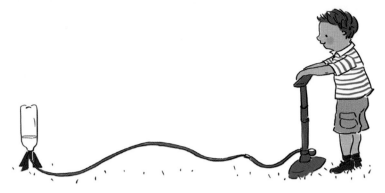

1 Ask an adult to help you. Following the instructions in the Rokit kit, put together the parts of the collar and attach the fins to it. Pour water into the bottle until it is about one-quarter full. Screw the collar onto the bottle. Attach the screw end of the yellow tube to the bicycle pump and push the brass plug on the other end into the hole in the black rubber nozzle in the collar.

2 Turn the rocket over so that it is standing on its fins. Stand it on some firm flat ground. Step back as far as you can—the hose is quite long—and begin pumping. You will see air bubbling up through the water and when the pressure inside the bottle is high enough, the rocket will blast off! If you are too close, you will get showered with water!

3 How high did the rocket go? This is hard to measure but you could time how long it stays in the air. How could you make the rocket go higher? Try adding a nose cone to reduce the air resistance on the way up or a parachute to slow it on the way down. Is a bigger bottle or a smaller bottle best? Does more or less water work best? Investigate and have fun.

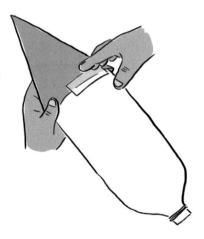

Behind the science
When you pump air into the bottle the air pressure inside rises. The pressure keeps rising until it is strong enough to force the hose out of the rubber nozzle and all the water behind it. The force of the water pushing down produces thrust, pushing up so the rocket shoots up against gravity. After a few seconds the thrust gets weaker until gravity is as strong as the thrust. This will be the rocket's highest point before gravity pulls it back down to Earth.

Chapter 8
Light and Energy

- -

THE QUESTIONS THIS CHAPTER ANSWERS:
Why is the Sun hot?
Why is the sky blue?
How far away are the stars?
Why are some stars different colors?

- -

What is light?

If you lie out in the Sun, you get hot, you are dazzled by its light, and, if you lie there too long, you get sunburnt! That's all because the Sun is a star where gas is always exploding and sending huge amounts of energy out into space. Visible light, heat (called infra-red light), and ultra-violet light (which gives you sunburn) are all types of energy from the Sun called electromagnetic radiation. Other types are radio waves, microwaves, Xrays, and gamma rays, which come from stars of different types, ages, and sizes. We all know their names because engineers have learned to use them in our TVs and radios, in microwave ovens, to look at broken bones, for wildlife cameras, and for many other things!

How hot is it today?

It is electromagnetic radiation from the Sun (mostly visible light and infra-red light) which warms Earth (and other planets). Summer days are warmer than winter days and countries closer to the equator are hotter than countries further from it because of the amount of radiation received from the Sun. More radiation reaches Earth in places where the Sun is high in the sky because it is not spread over a wide area (see page 38). However, there are lots of other things that control our weather and temperature from day to day including ocean currents, wind, and clouds. Earth reflects a lot of the radiation it receives back into space. Greenhouse gases can reflect radiation back down to Earth again, which is why global warming is becoming a problem (see page 71).

Wavelength and frequency

Visible light (the light you can see) and all these other forms of electromagnetic radiation travel through space in waves. Different types of radiation have different wavelengths. To understand this, think about waves in the ocean. Wavelength is the distance from the top of one wave to the top of the next. More waves will break on the beach the closer they are together. This is called the frequency of a wave. Short wavelength radiation has higher frequency and carries more energy.

Visible light and color

We can see the Sun and stars because they are light sources—they make their own light. When we see anything that is not a light source, what we see is light reflecting off it into our eyes. We can only see the Moon and the planets because they reflect sunlight.

What we call white light is made up of all the colors of the rainbow mixed together. Each color has a different wavelength. Violet light has the shortest wavelength of visible light and red, the longest. You can see the colors when they are separated into a spectrum. That happens when light moves from air into glass or water and gets bent a little. It bends because it slows down a tiny amount. The different wavelengths bend by slightly different amounts and spread out away from one another.

When white light hits an object some colors are absorbed (go into it) and some are reflected. We only see the color that is reflected back to our eyes. Uranus is a blue-green color because methane in Uranus's atmosphere absorbs red light but reflects blues and greens. Anything white reflects all the colors, while anything black absorbs them all.

A prism breaks up white light into a rainbow showing the seven colors of the visible spectrum.

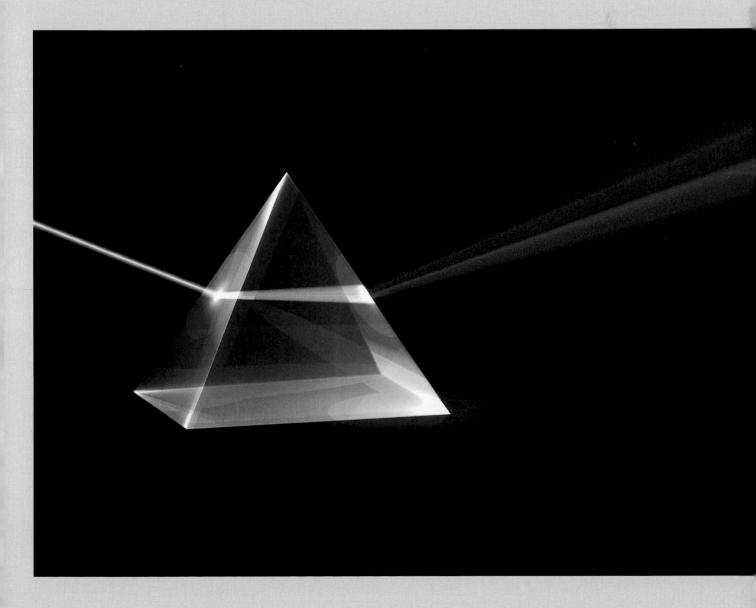

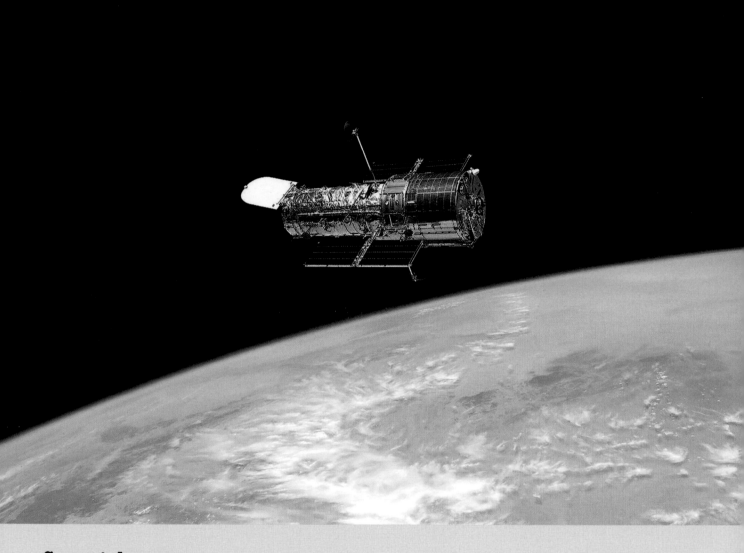

Super telescopes

We wouldn't know much about the universe without telescopes. After Galileo (see page 123) first spotted the moons of Jupiter in 1610, telescopes have been getting bigger and better year by year. Now there are many telescopes on satellites orbiting Earth; some like the famous Hubble Space Telescope look at visible light, others pick up the other kinds of radiation from space. Some of these telescopes work together to become even more powerful. Telescopes in space see clearer images because the waves do not get bent or scattered by Earth's atmosphere. Optical telescopes on

The Hubble Space Telescope can look far into space as it orbits Earth.

Earth are usually at the top of mountains where the air is thinner and drier (see page 70) and the light gets scattered less. The world's most powerful optical telescope is on top of a mountain in a desert in Chile and is simply called the VLT, or Very Large Telescope. Engineers are building another one now, also in Chile, called the ELT. Guess what that stands for!

The different colors of stars show how hot they are.

Why is the sky blue?

It is light bouncing off nitrogen molecules in the air that makes the sky blue. Shorter wave blue and violet light waves get scattered in all directions, whereas the other colors pass straight through the atmosphere, so more blue light reaches our eyes making the sky look blue. At sunset, when the light has further to travel, the less scattered red wavelengths are more likely to reach your eyes, making the sky turn red. On other planets the sky will be a different color because the different gases that make up their atmospheres scatter light in different ways. On both Mercury and the Moon the sky looks black because there is no gas to scatter the light.

Why are stars different colors?

Stars twinkle because they are so far away that not much of their light reaches Earth. The light that does reach Earth is bent and bounced around as it passes through the different layers of the atmosphere (see page 70). However, look at stars carefully, with binoculars if you have them, and you can see that some of them are slightly different colors. Because stars are light sources, their colors depend on how hot they are. The hottest ones are blue or bluish white; cooler ones are slightly red; our Sun is in the middle and appears yellow (but remember never to look directly at the Sun).

Star temperatures and their colors

Red: up to 7,232°F (4,000°C)
Orange: 9,032°F (5,000°C)
Yellow: 9,932–14,432°F (5,500–8,000°C)
White: 14,432–18,032°F (8,000–10,000°C)
Bluish-white: 18,032–36,032°F
(10,000–20,000°C)
Blue: 36,032–90,032°F (20,000–50,000°C)

The speed of light

Light travels in straight lines at 299,792 kilometers per second (186,282 miles per second). Albert Einstein (see page 123) worked out that nothing can travel faster than the speed of light. When astronomers talk about big distances in space, they measure them using light years. One light year is how far light can travel in one year. That is a long, long, long way. Remember, the Sun's light only takes eight minutes to reach Earth. It takes about four hours to reach Neptune, the farthest planet in the Solar System. That means light can go six times farther than Neptune in one day. Multiply that by 365 for the number of days in a year and you have a light year! Our nearest star, Alpha Centauri, is 4.35 light years away from us. When we look at stars, we are seeing what they used to look like. If something changes on Alpha Centauri, we would only see that change when the light reaches us 4.35 years later. Even when we look at another person on Earth, there's an incredibly tiny delay before the light that bounces off them reaches our eyes, but it's far too short to notice.

A Simple Spectroscope

There are lots of ways to make rainbows. Put your finger over a garden hose on a sunny day so you get a fine spray, and a rainbow will appear; stand your grandma's cut-glass vase on a sunny table and you'll see one somewhere on a wall; or look at the shiny side of an old CD. What you see is the visible light spectrum of sunlight. The colors are red, orange, yellow, green, blue, indigo, and violet, although nobody

ever really knows what indigo and violet look like and if you paint a rainbow, it's easier just to use purple!

You can remember the colors from their first letters with the mnemonic: Richard Of York Gave Battle In Vain or from the nonsense words ROY GBIV.

To discover more about distant stars and planets, astronomers use spectroscopes that break up their light into spectrums of different colors. Each star or planet has different bands of color in its spectrum. From these, astronomers can find out information about them such as which chemical elements are there, how hot they are, and how fast they are turning.

This is how to make a simple spectroscope.

1 Photocopy the template at the back of the book and stick it onto the card.

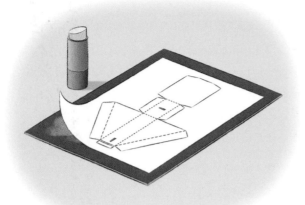

You will need

An adult to help

Copy of the template (see page 124)

Some thin card (black is best)

Glue stick

Pair of strong scissors

Craft knife

CD or DVD you no longer want

Masking tape

2 Score gently along the dotted lines on the template with the point of the scissors so that they are easy to fold. Cut out the outer shape. Fold the short dotted lines toward the top side of the template, and the longer dotted lines toward the underside of the template.

3 Ask an adult to help you use a craft knife to cut out the shaded boxes on the template.

4 Using masking tape or glue, join the small flaps from the middle section to the large triangle sides to make a triangle-shaped box. Fold over the large piece at the top and join to the largest flaps on the edges of the triangles to close the box, leaving a gap at the top. Make sure no light can get in through the joins.

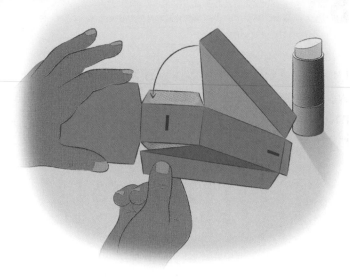

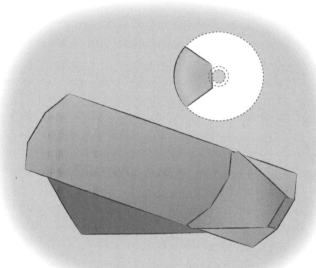

5 Ask an adult to help you cut out an arc-shaped section of the CD or DVD, which is about 3 inches (8cm) wide at its widest point. Strong scissors will cut through it. Slide the CD piece with the shiny side facing up toward the slit in the spectroscope. Fold down the small rectangle tab and stick it to the CD to hold in place. Apply more masking tape on the edges of the CD if you find light getting in.

6 Point the horizontal slit at different light sources BUT NEVER POINT IT DIRECTLY AT THE SUN. Look into the vertical slit and you should see a clear spectrum. Try candlelight (ask an adult) and different types of lamps. Can you see any differences in the colored bands?

Sunscreen Handprints

Ultra-violet light is not part of the visible light spectrum. It has a slightly shorter wavelength than violet light. We know it's there because it burns our skin and makes the color of paper or fabric fade. Sunscreen protects us if we use it properly. Here's an easy way to show it really does stop ultra-violet light reaching our skin.

You will need

A sunny day!

Sunscreen—factor 30
or higher

Sheet of colored
construction paper
(sugar paper)

Bricks or large stones to
weigh down the paper

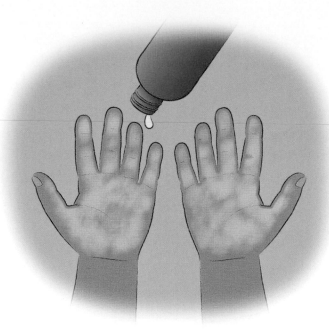

1 Smear sunscreen all over your hands.
Don't rub it in.

2 Press your hands onto the paper to make
prints of sunscreen. Repeat with more sunscreen
to make hand patterns all over the paper.

3 Put the paper in a sunny place and use bricks or large stones to weigh down the corners and stop it blowing away. Leave it for several hours.

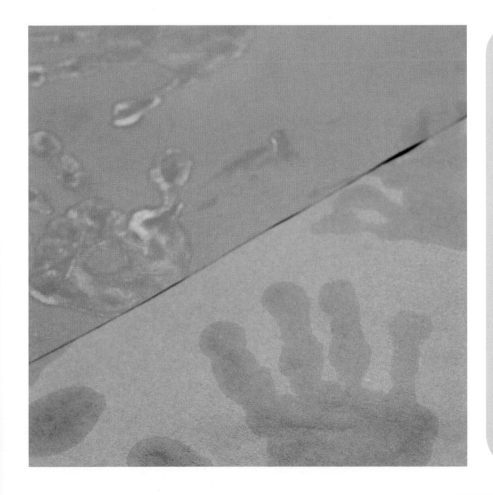

Behind the science

Ultra-violet light is harmful to living things, but the ozone layer stops most of it from reaching Earth. Some still does, and in this activity, the sunscreen stops ultra-violet light from reaching the paper, so the paper under your handprints stays bright. Everywhere else, the ultra-violet light makes the paper fade. Try it with ordinary moisturizing lotion to prove it is the sunscreen that is protecting the paper! This should make you believe in sunscreen and rub it on thickly whenever you are told to!

Cook a Marshmallow with Sunlight

You probably think of a telescope as a tube you hold up to your eye, but modern astronomers use reflecting telescopes. These are big, curved mirrors that collect light from a wide area and bring it together at a point called the focal point where there is a camera or another mirror to send it to the astronomer's eyes. Radio telescopes, which collect other types of electromagnetic radiation, also use curved receivers—if you have satellite TV, you will have one on your roof. In front of the dish, you will see a small object stuck out on an arm right in the center. This is the feedhorn and it is at the focal point of the dish. The radio waves from all round the dish are reflected onto this one point and then travel down a wire to your TV.

You can see how this works by cooking a marshmallow!

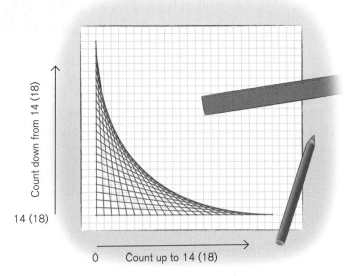

You will need

A sunny day!

Paper (½-inch/1-cm square paper is useful)

Sharp pencil

Ruler

Sharp scissors

Tissue box or similar (9 x 4 x 3 inches/23 x 11 x 7cm)

Thin card (letter size/A4)

Sticky tape (optional)

Glue stick

Aluminum foil

Marshmallow

Bamboo skewer

1 Be sure to keep to inches or centimeters, and don't mix them up for this activity. To make a parabolic reflector you need to draw a parabola. This is fun to do. First make a template by drawing two lines on a piece of paper, each 7 inches (18cm) long, to create a right-angle. Starting at the corner where the two lines meet, mark every ½ inch (1cm) along both lines (or use the squares marked on the paper). Mark the corner zero and then number along the bottom, counting up from 0 to 14 (or 0 to 18 if using centimeters). Number up the side, counting down from 14 to 0 (or 18 to 0 if using centimeters). Now draw a line to join number one with number one, two with two, and so on. As you join up more and more numbers you will see a curve appearing from the straight lines. This is a 2D parabola. 3D ones are more like the top of an eggshell, but they are very hard to make. When you have finished, cut carefully along the curve and sides to make a template.

2 Place the template on one of the long sides of the tissue box with the center of the curve in the center of the box and draw along the inside of the curve. Turn the box over and do the same on the other side, making sure the two curves are exactly on top of each another.

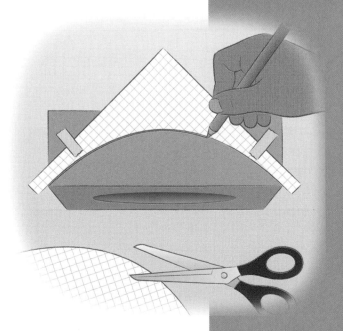

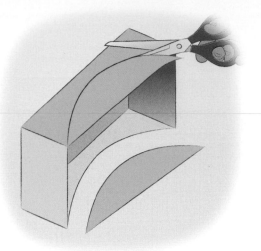

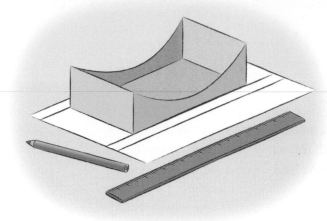

3 Cut off the whole front of the tissue box and cut along the curves at the top and the bottom.

4 Place the box bottom side down about ½ inch (1cm) from the edge of the piece of card and use a ruler to draw a line along the side of the box. Draw another line on the other side. Now use the ruler to continue the lines to the edges of the card. Draw one more line about a half inch (one centimeter) above the line in the middle. Cut along this line.

5 Score along the other two lines with the point of your scissors, so they will fold easily. Snip up to each line from the edge of the card, about every inch or so, to make lots of little flaps.

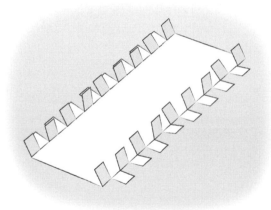

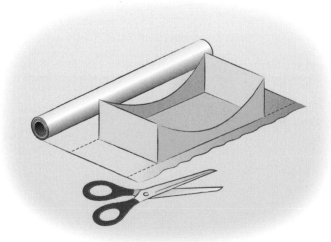

6 Use the back of the box as a template to cut out a piece of foil the same width as the box but longer to cover the piece of card.

7 Glue the piece of foil to the card, keeping it as smooth as you can because any wrinkles will reflect light all over the place and you won't have such a strong focal point.

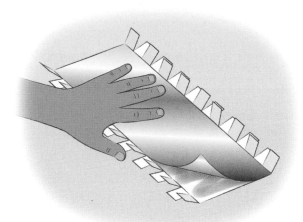

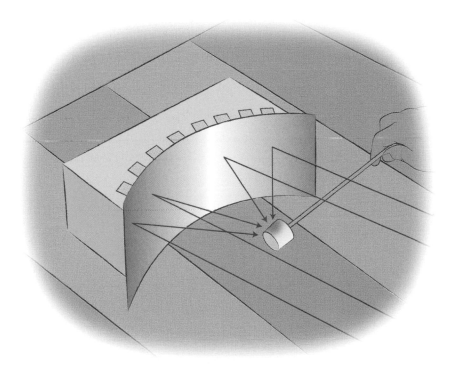

8 Carefully fit the foil-covered card into the curve of your box. Push one flap inside the box, then the next one outside, and so on, to make it fit well. Glue or use tape to hold the outside flaps in place. You now have a 2D parabola that's been made wider. This still has a focal point, although not such a strong one as a proper dish-shaped parabolic reflector would have.

9 Take your parabolic reflector outside into the Sun. Point it toward the Sun and see how the light is reflected off the foil onto one bright point in the center. This is the focal point. Do you remember that heat rays called infra-red rays also come from the Sun (see page 92)? They have a slightly longer wavelength than red light and are invisible. They are also reflected onto the focal point, so all the heat that hits the foil comes together at this point. Skewer the marshmallow, hold it at the focal point, and watch it melt. Enjoy your gooey treat!

Chapter 9
Space and Galaxies

THE QUESTIONS THIS CHAPTER ANSWERS:
How many stars can you see?
Can you see patterns of stars in the sky?
Why is there a long patch of mistiness across the sky at night?
Why are some stars brighter than others?

Galaxies

Our Solar System seems enormous to us, but it is nothing compared to a galaxy. A galaxy is made of dust, gas, and stars (together with their solar systems), all held together by gravity and spinning through space at enormous speeds. Our Solar System is part of the Milky Way galaxy, which contains over 200 billion stars and enough dust and gas to make billions more. At its center is an enormous black hole. It is a spiral galaxy, which means that most of its stars are in a fairly flat disk, with curved arms coming from the center. Our Solar System is in the middle of the disk on a small arm called the Orion Arm. Look up at the sky on a clear dark night and you will see the disk of the Milky Way as a faint band of misty light stretching from horizon to horizon. Use binoculars and you may be able to see the stars within it.

More than 25 years ago astronomers directed the Hubble Space Telescope to look at a very small, dark area of space which didn't seem to contain many stars. (The area could be covered by a pinhead if you held the pin up to the sky with your arm stretched out.) They kept the telescope looking at that same place for 12 days, collecting as much light as possible from deep in space. In that one small area, they picked up light from 10,000 galaxies of many different shapes and sizes! Because of that, astronomers now think there may be more than 100 billion galaxies in the universe. It is impossible to get your head around a number this big. They have also learned a lot about how different types of galaxies form and change through time.

The Big Bang

The universe is everything—all stars, planets, and space. The question "What is outside the universe?" can't be asked because that is all there is. But scientists know that the universe did not exist once. Astronomers watching stars realized that the whole universe is growing at enormous speeds. They measured how fast it is growing and then used math to work backward to try to understand what had happened before. They found out that the universe was very tightly squashed together until 13.8 billion years ago when it suddenly exploded and kept on expanding. That moment has been called the Big Bang and it involved unimaginable amounts of heat and energy. As the universe expanded, it cooled. In the first millionths of a second after the Big Bang, particles formed that are even smaller than atoms. Thousands of years later, these particles began to join together and make atoms. Millions more years passed and the atoms began to group into stars, then stars grouped into galaxies; dust formed into asteroids; asteroids joined to make planets; more stars were born and died. And the universe is still growing and changing. The sudden explosion has been called the Big Bang, but most scientists don't try to explain what exactly this was, only what happened after it.

Some of the galaxies photographed by the Hubble telescope are over 12 billion light years away. This means that their light has taken that long to reach us and so we are looking back in time to not very long after the Big Bang. That has given astronomers a lot to think about!

A nebula is a cloud of dust and gas from which stars are born.

The life cycle of a star

Throughout the universe, stars are born and die. A star begins in a huge cloud of dust and gas called a nebula. A clump of this comes together and begins to grow as gravity pulls in more and more dust and gas. The core becomes denser and denser and hotter and hotter until, when it is hot enough, a process called nuclear fusion begins, turning hydrogen gas into helium gas and sending massive amounts of energy into space. A middle-sized star like our Sun will stay at this stage, as a yellow dwarf star, for about 10 billion years before it runs out of fuel and begins to cool into a red giant. The Sun has about 5 billion years left before this happens and then it will become so big that its edges will reach Earth. It will eventually throw off its outer layers and shrink to the size of its core. It will become a white dwarf, smaller than Earth, but, because it will still have a huge mass, its gravity will remain very strong.

The life cycle of a really big star is different. It won't last as long as a small star because it burns through its hydrogen fuel more quickly.

It ends with a massive explosion called a supernova, which is so bright it can appear brighter than all the other stars in its galaxy. These are what can become black holes (see page 84).

Whatever way a star ends, it throws back dust and gas into space, which means more stars and planets can be born. But the stuff stars throw out has been changed by what has happened inside them. Stars begin as hydrogen and helium but, inside the core, atoms join together to make bigger atoms like carbon, oxygen, nitrogen, iron, and gold, and all the other heavier natural elements that make everything we know on Earth. We are all made of star dust!

When astronomers look at stars, just from their brightness and color, they can work out how much mass they have, how old they are, how they are changing, what elements they contain, how far away they are, and how fast they are moving.

Part of Orion

Part of Ursa Major/the Great Bear
(the Big Dipper or Plough)

Constellation Cups

If you are interested in astronomy, you will want to look at stars. The stars you can see easily from Earth are all in our Milky Way galaxy.

When you look at the stars, they seem to form shapes like a dot-to-dot pattern. The groups are called constellations. People have always noticed these shapes and given them names. Each of the twelve signs of the zodiac is a name of a constellation of stars, and there are many more. We see these stars as a two-dimensional pattern but, of course, they are in the three dimensions of space and some are much farther away than others. The ones that you can spot most easily are closer or brighter. Some are so faint that it is hard to see them. One way to learn the constellations is to make these constellation cups to project the stars as points of light onto a wall.

You will need

7 paper cups (base diameter about 2 inches/5cm)

Sharp scissors

Sticky tape

Large darning needle

Flashlight (torch)—one with a single bulb works better than one with multiple LEDs

Glue stick

Aluminum foil

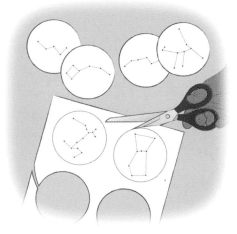

1 Photocopy the six constellation patterns on page 125. Cut out the circles. These are the templates for your star patterns.

2 Turn one of the cups over and place one star pattern on the base of the cup. Hold the pattern in place with a little sticky tape so it doesn't move.

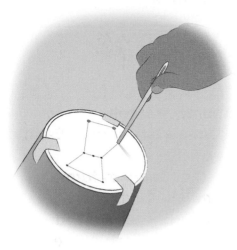

3 Use the needle to push through each of the stars marked on the paper so you make holes in the base of the cup. Be careful not to push the needle into your finger. Use something hard to push with if you can't get the needle through.

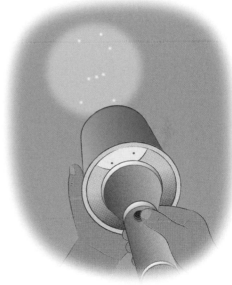

4 Remove the paper template and write the name of the constellation on the side of the cup. Repeat for the other five constellations using different cups. Now hold the flashlight inside one of the cups in a dark room and point it at the wall. You will see the constellation pattern on the wall. Have a go with the other cups to see some different constellation patterns.

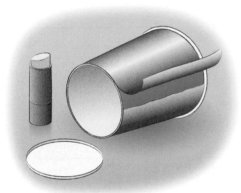

5 You can make the stars sharper by making a sleeve for your constellation cups. Carefully cut out the base of the last cup. Spread the outside of the cup with glue and stick a layer of foil all over it.

6 Slide one of the constellation cups inside the sleeve and any light coming through the sides of the cup will be blocked. You can use the same sleeve for each of your cups. You could make some more templates for constellations that are not included in the book—why not have a go with your own star sign.

Find the North Star

A good place to begin star watching is a visit to a planetarium. A planetarium is a big dome with stars and planets projected onto it which makes it seem as if you are really looking at the night sky on a perfectly clear night, with super-powered eyesight. During a show, experts will point out stars and planets and the groups of stars called constellations. They will also tell you a lot more about space in a fun way. It is easier to spot things in the night sky when you already have a good idea of what to look for!

The best way to see real stars is to go out on a cloudless night when there is a new moon (see page 49). The farther away you are from any artificial light sources, like streetlights or light from buildings, the more stars you will be able to see. You will be able to see the Milky Way as well as constellations and the North Star. You can download phone apps to help find the constellations and stars you are looking for.

Stars seem to move across the sky, but as you now know, that is just Earth rotating on its axis. For anyone living in the northern hemisphere the stars all seem to rotate around the North Star (which is also known as the Pole Star or Polaris) from east to west. If you live in the southern hemisphere, stars rotate around the Southern Cross and you will see different constellations.

The North Star is exactly above the North Pole, which is why it doesn't move. If you stood at the North Pole, it would be right above your head. If you stood on the Equator, it would be just above the horizon. The North Star was very important in the past because its height above the horizon told sailors their latitude. (Lines of latitude are imaginary lines drawn around the globe from east to west. They are parallel to the Equator.) This knowledge helped sailors navigate across the oceans. The North Star is not a bright star, so you need to learn how to find it.

1 Look north to find the Big Dipper. It could be in the north east or north west, but it is always very bright. Remember it may not be the same way up as you expect.

North Star

2 Find the two stars at the back of the chariot of the Big Dipper. These are called Dubhe and Merak. Then draw an imaginary line up from the chariot until you reach the next brightest star. That is the North Star, or Polaris.

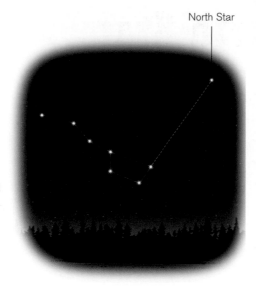

3 Polaris is one of the stars that makes up the constellation Ursa Minor (the Little Bear), also known as the Little Dipper. When you have found Polaris see if you can spot Ursa Minor next.

You will need

An adult to take you star watching

Warm clothes (if it is cold)

Flashlight (torch) to help you find your way safely to your star-gazing place and back home

Binoculars or a telescope (optional)

4 Can you find the other constellations you made with your constellation cups (see page 108)? Is the Milky Way visible? If you watch for just half an hour, you will notice the positions of the constellations change as you rotate around the North Pole on the spinning Earth.

Nebula in a Jar

The photos taken of nebulae are breathtakingly beautiful. A nebula is a cloud of dust and gas from which stars are born. These vast clouds of space dust and gas shine with different colored light depending on what they are made of. Sometimes the colors in the photos are not what you can actually see because the light that shines from nebulae isn't part of the visible spectrum (see page 93). The scientists use color to make their invisible light visible. This nebula in a jar is a simple way to recreate some of the magic of those images.

1 Have a look at some images of nebulae on the internet. Decide on the colors you will use to create yours. Use a separate cup for each color. Pour about half a cup of water (120ml) into each cup and add some food coloring. Mix the food colorings to make the colors you wish to use. Make some colors stronger than others.

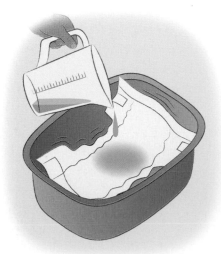

2 Spread out the diaper in the large container and pour over 2–3 cups (500–700ml) of water. Leave for a few seconds for the diaper to soak up the water.

3 Gently tear away the top layer of the diaper to reveal the loose, lumpy material inside. This is the absorbent gel which you need to make the nebula. Scrape the gel off the diaper into the container and add another cup or two of water. Watch the gel swell and become more like pieces of clear jelly.

You will need

About 5 cups for mixing

Measuring cup

Food colorings

Spoon

Disposable diaper (nappy)

Large container

Large glass jar with a lid (a Mason or Kilner jar works well)

Glitter

Star confetti

Bamboo skewer

4 Spoon some of the jelly into the glass jar to make a layer at the bottom. The thickness of the layer depends on how much you want of the first color. Don't flatten the jelly down—leave it as mounds and waves.

5 Pour over the first color—it will be absorbed by the jelly. Sprinkle some glitter and star confetti around the edges. These won't show in the middle, so you don't need any there. Push them down against the glass with the skewer.

6 Keep layering the jelly and different colors in this way, adding glitter and stars as you go. Use the spoon to push down between the layers in some places, so they don't look even but spread into one another. Fill the jar to the top and put on the lid. Stand the jar in the Sun to see it at its best.

Chapter 10
Life on Earth and Elsewhere

- -

THE QUESTIONS THIS CHAPTER ANSWERS:
Do aliens exist?
Why is there life on planet Earth?

- -

Is there life on Mars?

Everyone enjoys stories about aliens from outer space, even though many of them are pretty scary. So how many aliens do you think there really are in the universe? That is a very difficult question to answer.

So far astronomers have found no aliens anywhere in our Solar System. The latest Mars Explorer, Perseverance, is looking for signs that life might have existed in the past and may even still exist underground. But the kind of life they are looking for is microscopically small—perhaps something like bacteria or maybe like the slime you find at the side of a pond (but now fossilized). Living things look different to non-living things because they are made up of cells. Cells are what scientists are looking for on Mars.

What is evolution?

Life on Earth would have begun with something like bacteria, which are made of single cells, but over billions of years it evolved. Evolution is another difficult idea to understand. It's all about how slight changes in a living thing may make it better able to survive. If it survives, it will reproduce, and the next generations may have that same slight change which will help them survive too. Different groups of the same species change in different ways and branch off to become new species. Evolving life is like a tree that keeps on branching, but it is all very gradual. The scientist Richard Dawkins came up with a good idea to explain this. It's what's called a thought experiment.

Imagine you have a printed photo of your mother (or father). Now imagine finding another one of her mother, your grandmother, and another of her mother, your great grandmother, then your great, great grandmother. Keep getting more photos going back into the past (even though there were no photos back then). Place each photo one behind the other on a shelf. Keep going until your 185,000-greats grandmother, which would be at the end of a shelf of photos about 5 kilometers (3 miles) long. What would she look like? . . . A fish!

All of us have fish as our ancestors and, millions of years before that, slime. The important thing is that if you take any photo off the shelf and look at the one before it and the one after, you wouldn't see any difference. The changes from one generation to the next are so slight we don't notice them, just as we don't notice changes to our faces as we get older each day. But tiny changes over millions of years have meant life on Earth has evolved from slime to all the different species of plants and animals and other living things we see around us, including ourselves, humans.

What's so special about Earth?

There are lots of reasons why life was able to evolve into so many different species on Earth but not on Mars. Most of these have already been mentioned in the book.

1 Earth's orbit is in what is called the Goldilocks Zone. It isn't so close to the Sun that it is too hot, nor so far away that it's too cold. Like baby bear's porridge, it's just right. Earth's temperature means that there is liquid water here. Other planets have water, but the water is so hot it has turned to gas or so cold that it is ice. Life needs liquid water.

2 Earth has enough gravity to hold on to an atmosphere and a magnetic field that has kept the solar wind from stripping it away. The atmosphere acts like a quilt that keeps Earth warm and stops harmful ultra-violet rays from the Sun reaching the surface.

3 The gases that make up our atmosphere are all incredibly important for life. Nitrogen is one of the main building blocks of life; almost all life needs oxygen; plants need carbon dioxide to make food; and water vapor turns to rain to bring water from the oceans to land.

4 From its beginning, Earth had all the chemical elements necessary to build life, but there was no oxygen gas in the atmosphere. Then, by a lucky chance, a kind of slime evolved that could make its own food. It used a green chemical (called chlorophyll), water, carbon dioxide, and the energy from sunlight. It used some of the oxygen that is joined to the carbon in carbon dioxide (CO_2) and more that is joined to the hydrogen in water (H_2O), but it didn't need it all and what was left went into the atmosphere as gas.

Over millions of years, oxygen built up in the atmosphere. Green plants evolved from the slime and this amazing process, which is called photosynthesis, still keeps oxygen at a level of about 20 percent of Earth's atmosphere, which means we can breathe. The oxygen also gets dissolved in water and so living things in oceans, rivers, and other watery places use it too.

Photosynthesis equation

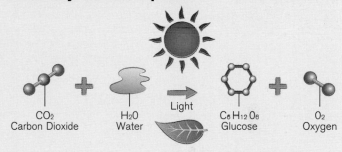

CO_2 Carbon Dioxide + H_2O Water → (Light) $C_6H_{12}O_6$ Glucose + O_2 Oxygen

5 Earth turns on its axis once every 24 hours. Only the Arctic and Antarctic are ever dark for a long time. If nights are too long, plants can't photosynthesize and without plants animals can't exist either because all animals need plants for food (even carnivores because carnivores eat herbivores).

6 Earth has the Moon orbiting around it. The Moon's gravity stops Earth wobbling on its axis and tilting over more. A different tilt would mean a change in seasons and climate. A climate that hasn't changed too much over millions of years has allowed life to evolve into millions of different species.

7 Life on Earth hasn't been wiped out by a huge collision with an asteroid, although big ones have caused mass extinctions, like the end of the dinosaurs. The planet Jupiter is a fairly close neighbor and this may have helped protect us.

Life might exist on Jupiter's moon Europa.

Is there life on other moons?

The next place in our Solar System where scientists think life could exist is on some of the moons of Jupiter and Saturn. Jupiter's moon, Europa, probably has an ocean of salty water beneath an icy crust and that might possibly support life. Saturn's moon, Titan, is even more promising. It is larger than Mercury, it has an atmosphere which is mostly nitrogen, like Earth's, and it has clouds, rivers, lakes, and seas but these are made of liquid methane and other chemicals, not water. Underground, there are lakes of water where there might just be life, or there could even be a different kind of life on its surface which doesn't need water.

Another moon of Saturn, Enceladus, has volcanoes that spray out icy water from an underground ocean. Most of this icy water falls back to the surface as ice crystals so that Enceladus looks as if it is covered in snow. But spacecraft have been able to pick up samples of the spray and have found that it contains most of the ingredients needed for life. If we do find life on these moons, it will still only be the very simple cells, which are the beginnings of life.

Is there life on exoplanets?

Now astronomers are looking for life beyond our Solar System. They search for exoplanets (planets orbiting different stars in different solar systems) that might be able to support life. The problem is finding them when they are so far away. The light they reflect isn't easy to pick up on Earth against the brightness of the stars they orbit. They have to be spotted in clever ways.

Some astronomers look for a slight wobble in the rotation of a star on its axis. A wobble is caused by the pull of gravity between the star and a planet that orbits it. Big planets, like Jupiter, make a star wobble quite a lot (Jupiter wobbles our Sun), but small, Earth-like planets don't cause much wobble and so are difficult to spot.

Other astronomers look for a slight dimming of a star, which happens when a planet's orbit takes it between the star and Earth and blocks a little of the star's light. This is called a transit. How much the star is dimmed helps astronomers work out the size of the planet. They can also find out how much time there is between "transits" and from this they can work out how far the planet is from its star. They can work out if it is in the Goldilocks Zone and might be suitable for life.

A space telescope called Kepler has already found thousands of exoplanets and future missions will find thousands more orbiting our nearest stars. There are so many galaxies, each with so many stars in them, that there must be billions of planets in the universe.

As there are so many planets, there is a very good chance that life will have evolved on some of them—just like if someone buys millions of lottery tickets there's a good chance one will win! But what would that life be like? It might never have evolved beyond slime or it might be super intelligent and have even better technology than humans. It would probably not look like us. How can we find out? The nearest exo-planet that might be suitable for life orbits Alpha Centauri and is 4.2 light years away. The fastest spacecraft developed so far would take 80,000 years to reach it, so we can't visit!

How can we find out if there are aliens?

One way we can find out if there is anyone else out there is by listening for them. If aliens have developed technology somewhere else, they might be sending out radio messages. Astronomers have started to use huge parabolic radio telescopes (see page 102) to look for unusual patterns in the radio waves that come from space. These might just have been made deliberately rather than being part of the random signals of space.

Even if we do find aliens this way, there is another problem. When were the signals made? If they came from an exoplanet in a galaxy millions of light years away, they would have been made millions of years ago. Any aliens who made the signals may have long since become extinct.

But what if there are aliens who are also looking for signs of life on other planets? Two Voyager spacecraft, launched in 1972 and 1973, were designed to travel past the planets, taking photos and sending back information, and then to keep going beyond our Solar System. A gold plate was put into each of them, engraved with symbols that could communicate with alien species. These include pictures of a naked man and a naked woman, so that any aliens who find the plates will know what we look like. They are standing beside the Voyager spacecraft to show how tall we are. There is also a diagram of our Solar System and other mathematical symbols that might be understood by an alien who has advanced technology.

We have also begun sending radio signals into space just in case we are not the only ones listening out for unusual signals!

Our precious blue planet

Before astronauts first looked out of their space capsule window and saw the whole of Earth as a blue and white sphere in space, people didn't really understand how small and fragile Earth is. Now we know how easy it is to harm it. We have polluted our oceans and rivers, cut down our forests, hunted animals to extinction, destroyed habitats, and filled our precious atmosphere with greenhouse gases.

Our planet is very special. For 3.5 billion years, life has been evolving on Earth and that life is part of an ecosystem where everything depends on everything else. Human beings have only begun to mess it up in the last few hundred years. We need to learn to take care of it before it stops taking care of us.

Scavenger Hunt

Even in your backyard or local park you will find many different kinds of living things, all of which have evolved from that single-celled organism which began life 3.5 billion years ago. See how many you can find on this scavenger hunt and think about how strange and wonderful life on Earth really is. Then make sure you keep working to protect it.

1 Write the alphabet down the side of your paper.

You will need

Piece of paper

Pencil

Magnifying glass (optional)

Identification books or apps for plants and invertebrates (bugs)

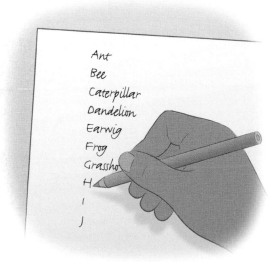

2 Now for each letter find at least one species of plant or animal (mammals, birds, bugs, etc.) beginning with that letter. Look in corners, under stones, and in the soil. Use identification books or apps to find the names of things you don't recognize. Can you make it right through the alphabet?

3 For the letter x use any species that has x in its name (like foxglove). If children all over the world, in millions of different habitats, did this same activity, just think how many different species would have been spotted!

The Planets

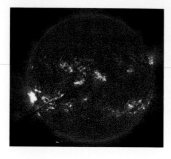

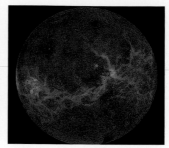
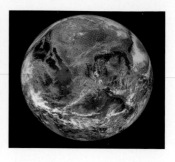

The Sun

The Sun is a huge sphere of glowing gases that sends out enormous amounts of energy as light, heat, and other types of electromagnetic radiation (see page 92). Everything in the Solar System orbits around the Sun, held there by its huge gravitational pull. It is a yellow dwarf star and is about halfway through its life. In another 4 to 5 billion years, it will become a red giant and then a white dwarf.

Light and heat from the Sun are essential for life, but solar wind also streams out from the Sun. This is made of charged particles and is dangerous to living things. Our atmosphere protects us from the solar wind, but it can damage satellites and other electronic equipment.

The Sun is incredibly hot at its core, cools toward its surface, but mysteriously starts to get hot again in its atmosphere, the temperature rising rapidly toward the outer layer, the corona. This layer can only be seen in a total solar eclipse and is where the solar winds begin.

The Sun in numbers
Diameter: 1,391,000km (864,000 miles). The Sun is 109 times wider than Earth. A million Earths could fit inside it.
Temperature at the core: 27 million °F (15 million °C)
Temperature in photosphere: 10,000°F (5,538°F)
Temperature in the corona: 2 million °F (1.1 million °C)
Time to rotate once on its axis: 27 days

Mercury

Mercury is the closest planet to the Sun and the smallest planet in the Solar System. It is a rocky planet and has lots of craters on its surface. It spins slowly on its axis, so each day is very long. It has a very fast orbit around the Sun, so its year is very short. Mercury doesn't have an atmosphere, just a thin exosphere made up of molecules blasted off the surface by meteors and the solar wind. Without an atmosphere to hold in the heat, it gets very cold on Mercury at night. The difference between night and day temperatures can be as much as 1000°F (538°F). Where there is a big difference, an average doesn't mean much, so for Mercury, the Moon, and Mars we have given the highest and lowest temperatures.

Mercury in numbers
Moons: 0
Rings: 0
Diameter: 4,879km (3,032 miles)
Distance from the Sun: 0.4 AU
Time to rotate once around its axis: 58 Earth days = one day on Mercury
Time to orbit the Sun: 88 Earth days = one year on Mercury
Time for light to reach it from the Sun: 3.2 minutes
Highest day time temperature: 840°F (449°C)
Lowest night time temperature: -275°F (-170°C)
Weight difference: If you weigh 66lb (30kg) on Earth, you would weigh 25lb (11.3kg) on Mercury.

Venus

Venus is the second planet from the Sun and another rocky planet. It has a very thick atmosphere made mostly of carbon dioxide and other greenhouse gases which trap the heat, so it is hot enough to melt lead on its surface. The atmosphere stops it getting colder at night. You can't see the surface of Venus because it is covered with a thick layer of cloud made of drops of sulfuric acid, but it has many volcanoes. The air pressure on Venus is so high that if you stood on its surface, you would be squashed (the same would happen if you tried to go 3,000ft/ 900m under the sea). Venus spins so slowly on its axis that a day is longer than its year, and because it rotates backward the Sun rises in the west and sets in the east.

Venus in numbers
Moons: 0
Rings: 0
Diameter: 12,104km (7,521 miles), which is a similar size to Earth
Distance from the Sun: 0.7 AU
Time to rotate once around its axis: 243 Earth days = one day on Venus
Time to orbit the Sun: 225 Earth days = one year on Venus
Time for light to reach it from the Sun: 6 minutes
Average temperature: 880°F (471°C)
Weight difference: If you weigh 66lb (30kg) on Earth, you would weigh 60lb (27.2kg) on Venus.

Earth

Earth is the third planet from the Sun and is a rocky planet. It is unique in our Solar System because it is mostly covered with liquid water. Oceans cover 70 percent of Earth's surface. It has an atmosphere that is 78 percent nitrogen and 21 percent oxygen with small amounts of argon, carbon dioxide, and methane.

Earth in numbers
Moons: 1
Rings: 0
Diameter: 12,756km (7,926 miles)
Distance from the Sun: 1 AU
Time to rotate once around its axis: 24 hours
Time to orbit the Sun: 365.2 days
Time for light to reach it from the Sun: 8.3 minutes
Average temperature: 61°F (16°C)
Highest temperature recorded on Earth: 136°F (58°C)
Lowest temperature: -126°F (-88°C)

The Moon

The Moon is a lot like the planet Mercury, but rather than orbiting the Sun it orbits Earth. Like Mercury, the Moon only has a thin exosphere rather than an atmosphere, so during its long nights (the length of 13½ Earth days) it gets very cold and during its day (the length of 13½ Earth days) it gets very hot. Astronauts who walked on the Moon had to wear spacesuits to protect them from heat, cold, harmful radiation, and low air pressure as well as to provide them with air to breathe.

The Moon in numbers
Diameter: 3,474km (2,159 miles), which is about a quarter the size of Earth
Distance from Earth to the Moon: 384,400km (238,855 miles)
Time to rotate once around its axis: 27 Earth days = one day on the Moon
Time to orbit Earth: 27 days
Highest day time temperature: 260°F (127°C)
Lowest night time temperature: -280°F (-173°C)
Weight difference: If you weigh 66lb (30kg) on Earth, you would weigh 11lb (4.9kg) on the Moon.

Mars

Mars is the fourth planet from the Sun and the last of the rocky planets. It is called the red planet because there is a lot of iron in its rock which turns to rust-colored dust. The biggest mountain in the Solar System, Olympus Mons, is on Mars. It is about two-and-a-half times taller than Everest.

Mars has a very thin atmosphere, so it is a cold, desert world. However, there is enough atmosphere to cause winds and dust storms. Mars was once much warmer and had rivers and seas. There is still water on Mars, but it is locked into ice caps at its poles or underground. Mars has two moons, Phobos and Deimos. Phobos is descending toward Mars, and scientists believe that one day it will crash into it.

Mars in numbers
Moons: 2
Rings: 0
Diameter: 6,792km (4,220 miles), which is about half the size of Earth
Distance from Sun: 1.5 AU
Time to rotate once around its axis: 25 Earth hours = one day on Mars
Time to orbit the Sun: 627 Earth days = one year on Mars
Time for light to reach it from the Sun: 12.6 minutes
Highest day time temperature: 70°F (20°C)
Lowest night time temperature: -225°F (-153°C)
Weight difference: If you weigh 66lb (30kg) on Earth, you would weigh 25lb (11.3kg) on Mars.

Jupiter

Jupiter is the fifth planet from the Sun and is the first of two gas giants. It is mostly made of hydrogen and helium gas, and has no surface, but if you traveled toward its center, the gas molecules would be pulled closer and closer together until they became a liquid and then, at the very center, a solid.

Jupiter is by far the largest planet in the Solar System. More than 1,500 Earths would fit inside Jupiter and it has almost twice the mass of all the other planets put together. If Jupiter had got about 80 times bigger when it was formed, it would have begun to burn and then it would have become a star. Because it is so massive it has the strongest gravitational pull of all the planets in the Solar System and, since it spins so fast on its axis, it also has the strongest magnetic field.

It is covered in stripes of thick, swirling cloud and has enormous storms like the Great Red Spot, which has been raging for hundreds of years.

Jupiter has many moons but the four most famous ones each has something special to remember them by:
Io is the most volcanic place in the Solar System.
Europa might have an ocean of salty water beneath an icy crust that might possibly support life.
Ganymede is bigger than Mercury and has its own magnetic field.
Calisto has more craters than anywhere else in the Solar System.

Jupiter in Numbers
Moons: 79
Rings: 3 faint ones
Diameter: 142,984km (88,846 miles), which is 11 times wider than Earth
Distance from the Sun: 5.2 AU
Time to rotate once around its axis: 10 Earth hours = one day on Jupiter
Time to orbit the Sun: 4,331 Earth days = one year on Jupiter
Time for light to reach it from the Sun: 43.2 minutes
Average temperature: -162°F (-108°C)
Weight difference: If you weigh 66lb (30kg) on Earth, you would weigh 160lb (75.2kg) on Jupiter, but you could never stand on Jupiter because it has no surface.

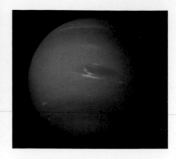

Saturn

The sixth planet from the Sun is Saturn. This is another gas giant but it's not as big as Jupiter. Saturn is similar to Jupiter as it is made mostly of hydrogen and helium gas. It is famous for its beautiful and spectacular rings, which are all made up of chunks of rock and ice orbiting the planet.

Like Jupiter, Saturn has many moons with two very special ones. Enceladus has volcanos which spray out icy water from its underground ocean. Titan is larger than Mercury and is another place where life might be found.

Saturn in numbers
Moons: 82
Rings: 7
Diameter: 120,536km (74,898 miles), which is nine times wider than Earth
Distance from the Sun: 9.6 AU
Time to rotate once around its axis: 11 Earth hours = one day on Saturn
Time to orbit the Sun: 10,747 Earth days = one year on Saturn
Time for light to reach it from the Sun: 79.3 minutes
Average temperature: -218°F (-138°C)
Weight difference: If you weigh 66lb (30kg) on Earth, you would weigh 70lb (31.9kg) on Saturn, but you could never stand on Saturn because it has no surface.

Uranus

Uranus is the seventh planet from the Sun and the first of two ice giants. It has a small, rocky core with a mantle made of a thick, icy soup of methane, ammonia, and water. It has an atmosphere of hydrogen and helium, like the gas giants, with some methane which makes it look blue. The other three giant planets have patterns made by storms, but photos show Uranus to be a smooth pale blue.

Uranus was only discovered in 1781. It is so dim that everyone thought it was a star before that. It seems that Uranus was in a mega-collision 3 to 4 billion years ago and was knocked almost on its side. Each of the poles gets 21 years of darkness followed by 21 years when the Sun never sets! It also rotates east to west like Venus.

Uranus in numbers
Moons: 27
Rings: 13 faint ones
Diameter: 51,118km (31,763 miles), which is four times wider than Earth
Distance from the Sun: 19.2 AU
Time to rotate once around its axis: 17 Earth hours = one day on Uranus
Time to orbit the Sun: 84 Earth years = one year on Uranus
Time for light to reach it from the Sun: 164 minutes
Average temperature: -320°F (-195°C)
Weight difference: If you weigh 66lb (30kg) on Earth, you would weigh 59lb (26.6kg) on Uranus, but you could never stand on Uranus as it has no surface.

Neptune

Neptune, the second ice giant, is the farthest planet from the Sun, 30 times farther from the Sun than Earth. That means the Sun would look 33 times smaller in the sky than it does on Earth, just like a very bright star. There is never much daylight on Neptune.

Neptune and Uranus are made of similar ice and gases, but Neptune behaves differently because it wasn't knocked over in a mega-collision, as was the case with Uranus. It is a cold, stormy world where winds blow clouds of frozen methane across the surface at 2,000 kilometers per hour (1,200 miles per hour). Astronomers have observed dark spots in its atmosphere, which are storms like Jupiter's Great Red Spot. Neptune should be much colder than Uranus because it is much further from the sun, but it's not. Scientists think this is because of heat coming from its rocky core. Uranus doesn't have this heat perhaps because of the collision which knocked it onto its side.

Neptune was only discovered because astronomers studying Uranus noticed something strange about its orbit. They decided that it was being pulled by the gravity of another planet and worked out, by math, where that planet should be. Sure enough, when another astronomer looked at that place with a telescope, he found Uranus. Its orbit is so huge that it has only completed one orbit since it was discovered in 1846!

Neptune in numbers
Moons: 14
Rings: 5 faint ones
Diameter: 49,528km (30,775 miles), which is four times wider than Earth
Distance from the Sun: 30 AU
Time to rotate once around its axis: 16 Earth hours = one day on Neptune
Time to orbit the Sun: 165 Earth years = one year on Neptune
Time for light to reach it from the Sun: 248 minutes
Average temperature: -331°F (-201°C)
Weight difference: If you weigh 66lb (30kg) on Earth, you would weigh 74lb (33.7kg) on Neptune, but you could never stand on Neptune because it has no surface.

Some Famous Astronomers

Nicolaus Copernicus was born 1473 in Poland. At that time everyone believed that Earth was the center of the universe with the Sun, the stars, and the planets all revolving around it, but the astronomers of the day couldn't agree on the order of the planets from Earth. Copernicus studied astronomy and realized that the reason they could not agree was because the planets revolved around the Sun. He worked out the correct order of planets from the Sun and how long each took to orbit it. He also worked out that Earth must revolve once on its axis each day and that it takes one year to orbit the Sun, but he did not put his revolutionary ideas into a book until the year of his death in 1543.

Galileo Galilei was born in 1564 near Pisa in Italy. He was an exceptional astronomer, mathematician, physicist, and thinker. He worked on problems to do with the speed at which objects fall (see page 86), pendulums and forces, and motion. In 1609, he heard about the invention of a telescope in the Netherlands and, without having seen it, set about making one. He built one which magnified things by 30 times and, with this, he first saw the mountains and valleys on the Moon, sunspots on the Sun, the four largest moons of Jupiter, Saturn's rings, and the phases of Venus (which are like the phases of the Moon). Because of his observations he realized that Copernicus was right about Earth and the planets revolving around the Sun, but this went against the belief of powerful teachers in the church. When he wrote a book about it, he was convicted of heresy (going against the church) and sentenced to house arrest for the rest of his life. He died in 1642.

Isaac Newton was born in England in 1643. He is still considered to be one of the greatest scientists who ever lived. His theory about gravity, called the Law of Universal Gravitation, made it possible to explain the movements of planets and the Sun. His three laws of motion are the foundation of the whole science of physics. He discovered that light could be split into the colors of the spectrum, and he invented the reflecting telescope used by astronomers today. He also invented a whole new branch of mathematics, now called calculus, which is extremely important in science and engineering. He died in London in 1727. He said once, "If I have seen further than others, it is by standing upon the shoulders of giants," showing that he recognized that all advances in science depend on the work of scientists who have gone before.

William and Caroline Herschel William Herschel was born in Germany in 1738 but spent most of his life in England. He first worked as a musician but then became interested in astronomy. Finding that the telescopes of the day were not powerful enough, he set about making better ones. His telescopes were the best of the time and with them he discovered that nebulae were not misty clouds but clusters of stars. He also observed a new object in the sky which turned out to be Uranus, and he discovered infra-red radiation. He and his sister Caroline worked together and she became an astronomer herself. She discovered several comets and was the first woman who was paid to be a scientist. William died in 1822 and Caroline in 1848.

Albert Einstein was born in Germany in 1879 but he later moved to America where he lived until he died in 1955. He is probably the most famous scientist of the twentieth century. He realized that although Isaac Newton's laws worked well, there were things in physics they couldn't explain completely, including gravity. He developed the Special and General Theories of Relativity which explain gravity through the idea of space–time (see page 89). These theories have helped astronomers understand how the whole universe works, including neutron stars and black holes. His most famous equation $E=mc^2$ stands for **E**nergy equals **m**ass times the speed of light (**c**) squared. In other words, mass can change into huge amounts of energy, which is what happens in the Sun and stars to produce electromagnetic radiation (see page 92).

Templates

A Simple Spectroscope

Pages 96–98

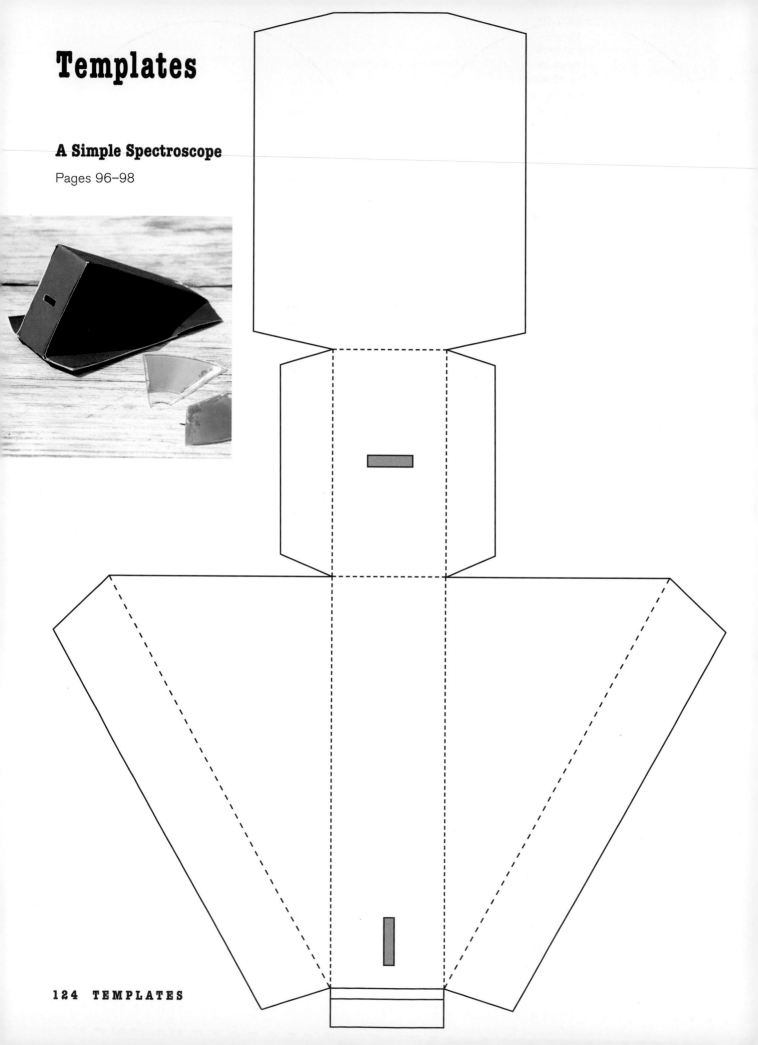

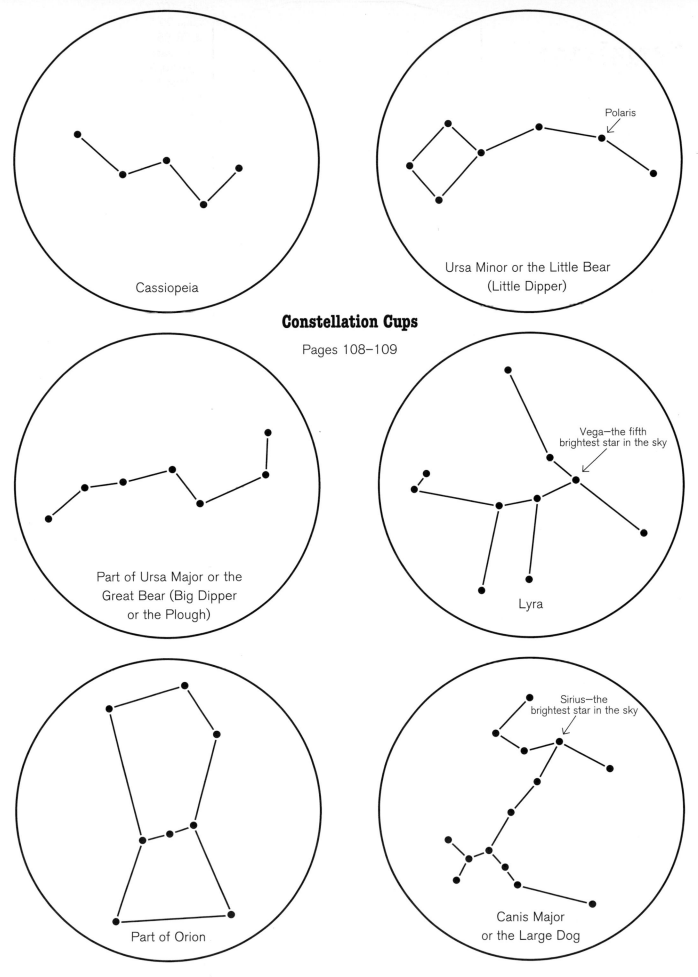

Cassiopeia

Ursa Minor or the Little Bear
(Little Dipper)

Polaris

Constellation Cups

Pages 108–109

Part of Ursa Major or the
Great Bear (Big Dipper
or the Plough)

Vega—the fifth
brightest star in the sky

Lyra

Part of Orion

Sirius—the
brightest star in the sky

Canis Major
or the Large Dog

Index

Index entries in bold are activities.

Resources

If you want to find out more about space and planets, go to this brilliant website designed especially for kids:
https://spaceplace.nasa.gov

To view amazing images of space, go to:
https://www.nasa.gov/multimedia/imagegallery/index.html

Picture credits

Alamy
page 85: Granger Historical Picture Archive/ Alamy Stock Photo

NASA/JPL-Caltech
pages 1, 2, 4, 10, 33, 35, 45, 46, 57, 69, 84 below, 94, 95, 107, 117, 120, 121 left and middle, 122 left and middle

Shutterstock.com
page 3: Alex Mit; 5: D1min; 6–7: barmalini; 9: Triff; 11: Skylines; 20: Rido; 21: Antonov Maxim; 27: fluidworkshop; 32: GraphicsRF.com; 47 and 81: kdshutterman; 54: Colin Hayes; 55: fboudrais; 56: Jamen Percy; 70: Designua; 71: Ziablik; 82: Castleski; 83: Vera Larina; 84 top: ChiccoDodiFC; 93: MicroOne; 111: Yuriy Mazur; 116 and 121 right: Dotted Yeti; 122 right: 24K-Production; Star graphics on pages 12, 15, 17, 22, 25, 37, 40, 59, 61, 65, 73, 75, 76, 79, 90, 97, 100, 103, 113, 119: Nikolaeva

page 114: "Thought experiment" is from *The Magic of Reality* by Richard Dawkins (Bantam Press, 2011).

Acknowledgments

Thanks to NASA's Space Place for great ideas and reliable information on everything to do with space.

Thanks also to all the many educators who share ideas for science experiments and activities on websites and YouTube.

Finally, my thanks to my editors at CICO Books, who managed to keep the book on track despite COVID lockdowns.